The Grass Is Greener

Till You Get To The Other Side

To Ted + Tanya
Dominus vobiscum
Tershiá

by Tershia Lambrechts

Acknowledgements

I should like to thank the following people whose contributions to the creation of this book are greatly appreciated:

- Melissa Beske for generously making her research into Domestic Violence in Belize available to me.
- Guadalupe Lima for allowing me to voice her opinion on business aspects.
- Jerry Larder for the use of his poem which amusingly sums up the Belize ex-pat experience.
- Mark Hazeldine, our fellow sojourner in Belize, for his constructive criticism!
- Hermann and Tracy Nell for allowing me the use of their tranquil cottage-on-the-lake where this book began.
- Pam Mann whose enthusiasm for the book and advice was most helpful.
- My friends Alice McDonald and Wendy Johnsen who have always been my support.
- My husband Hugo who believed in me and kept the pot boiling somewhat while I laboured at the computer.
- Finally, I thank the people of Belize for giving me a reason to write this book.

Suite 300 – 990 Fort Street
Victoria, BC, Canada V8V 3K2
www.friesenpress.com

First Edition — 2014

ISBN
978-1-4602-5326-7 (Hardcover)
978-1-4602-5327-4 (Paperback)
978-1-4602-5328-1 (eBook)

1. *Biography & Autobiography, Personal Memoirs*

Distributed to the trade by The Ingram Book Company

This book is dedicated to the women and children of Belize who are victims of domestic violence and sexual abuse.

Chapter 1

The world is a great book, of which they that never stir from home read only a page.

St. Augustine (354-430)

It was early January 2003 in Belize and the bus rattled along the Western Highway. Not a real highway as we are used to in North America, but a narrow road with many potholes and no shoulder. The seats were worn and uncomfortable and the air conditioner dripped water on the passengers seated directly below. Some passengers, more accustomed to a soft life, grumbled. That would not include us; my husband Hugo and I prided ourselves with being made of sterner stuff. However, this wasn't the air-conditioned tour bus all of us expected from the brochure, but an old recycled school bus from the United States; its original identity vaguely concealed by a new paint job.

We were an eclectic group of approximately twenty people; citizens from the United States along with Hugo and I from Canada, who signed up for a five day "Belize Relocation Conference and Tour" to explore the possibilities of retiring in the tropics. We all had our own reasons for contemplating such a move; from tax evasion to seeking adventure in a sunnier climate. We had been influenced by what we considered to be one of the most comprehensive books on the subject, *Belize Retirement Guide: How to live in a Tropical Paradise on $450 a Month* by Bill and Claire Gray[1], facilitators of the conference and tour.

Map of Belize

The book extolled the advantages and delights of living in tropical Belize. It included all the information one needed to know on how to make the actual move, and all the facts and advantages of daily living in the country. Some of the perks that drew our attention were: the low cost of living virtually tax-free; all the modern conveniences in one's home, which included high-speed internet, cable television, fibre-optic telephone lines and no heating bills; good postal service; unarmed police; low cost medical and dental care; abundant tropical fruits; beautiful affordable homes and powdery white beaches along the coast, all that on US $450 a month. We thought the amount sounded too good to be totally true, yet worth investigating. A published statement by the Belize Ministry of National Security and Immigration gave credence to this possibility by stating:

> *Belize's new "retiree law" makes this country's program the best in the world. On September 15, 1999 Belize's new retiree legislation became law - meaning that Belize now offers what we believe to be the most attractive incentive program anywhere for foreign retirees. If you're looking for tax-free living, put Belize at the top of your list.*
>
> *The law aims to encourage and promote the inflow of foreign capital into Belize by offering certain tax exemptions and incentives to Qualified Retired Persons. It's a win-win situation for you and for the Belize government.* [2]

The Belize Relocation Conference and Tour seemed an excellent way to find out all the relevant details and experience the country first hand. Therefore, we registered at the given California address and in January we flew from Vancouver to Belize City.

Upon our arrival at the Philip Goldson International Airport, we transferred to a small local island-hopping air service for the fifteen-minute flight to the town of San Pedro on Ambergris Cay (the main tourist destination and the largest of about 200 cays of various sizes, dotting the Caribbean Ocean).

The second longest reef in the world lies just beyond Ambergris Cay, creating calm and shallow waters between the reef and the mainland. We were booked into the Sunbreeze Hotel, charmingly situated on the beach. Our room was clean with basic furniture made from local hardwood: certainly adequate enough but nowhere near a three star rating. Our expectations had been a bit too high, but we were not going to let that deter our enthusiasm for this trip.

The first two days of the tour were taken up with lectures held on an upper level open patio where we were seduced by the view of the calm turquoise ocean and palm trees swaying in the breeze. Overhead the pelicans were gliding down to the shallow water's edge to feed on the discarded bits of the local fishermen's morning catch. It was a scene straight out of a tropical escape magazine.

Our speakers included representatives from various government departments such as: customs; immigration and tourism; business people; an expat lawyer who had forsaken Kauai for Belize; expat realtors and a few others who had advice to give. After two full days of imparted information and answers to our questions, we felt adequately informed on everything there was to know about retirement in Belize and now looked forward to day three: the tour itself.

By this time we had met our American expat tour guides Judy, Andy and Lynn who lived in the town of Corozal, situated on the Bay of Campeche just south of the Mexican border. On day three we clambered into more small inter-island planes and flew from San Pedro to Corozal. Upon landing on the dirt strip, the tour bus awaited us.

There was much to see and many stops to make in one day so the bus wasted no time in rumbling into the town of Corozal. Our first stop was at an intriguing local open-air market, where everyone shopped for fresh fruit and vegetables. One of the vendors ran an open-air butcher shop. The odour turned my stomach. I wondered if there were proper butcher shops with proper refrigeration and no flies in the country. The other fresh produce seemed good enough. We were especially taken by the abundance of fresh bananas: eight for one Belize dollar (fixed to the US dollar at a rate of 2:1).

As we had not as yet seen anything remotely like a 7Eleven corner store where we could buy snacks to eat along the way, we stocked up on a variety of unfamiliar small sweet bananas called *manzanitas*. After the market came a hurried stop at a hardware store in the centre of town. Not a Home Depot by any means, but it seemed to stock all the basic plumbing necessities one would require when building a house.

Corozal has an expat community of approximately 2000, including a retirement enclave known as Consejo Shores a few miles outside the town where we headed next. Consejo Shores did not appear to be fully developed and it certainly did not live up to the manicured image we had formed from the description in the tour brochure. However, we had already decided that living in an expat enclave was not for us. One often heard faint rumours of decadent living with much rum swilling and wife swapping in some expat communities. We preferred to live in the general community or better still, we aspired to owning our own little patch of jungle.

After a quick trip through Consejo Shores the bus trundled across the border and into the Mexican frontier town of Chetumal. The streets were wide and even though a little untidy looking, the many buildings and homes were painted in bright colours, creating a friendly casual atmosphere. We were hurried through an American-style supermarket in a shopping mall, with hardly enough time to have a thorough look at everything available on the shelves.

Then it was on to visit a small medical hospital and dental clinic; both seemed clean and well equipped. Speaking from experience, Andy assured us of the excellent medical and dental treatment one would receive (if necessary) and not too far from Belize either. It did not occur to us to enquire why one would need to go to Mexico for treatment if Belize offered their own low cost medical treatment as mentioned in the retirement guide.

The tour bus

Back across the border into Belize, we stopped at a pleasant home and garden situated on an acre overlooking a lagoon that belonged to Chuck and Judy, expats from Oregon State. They had built their house on the foundation of a half-finished abandoned building: of which there are many in Belize. The reasons for these failed building projects are the result of people with ambitions that outweigh their financial realities. Chuck, being in the building business, gave us good information on the "dos and don'ts" when undertaking house construction in the country. We would come to rely on them for helpful advice.

After a long day of being shaken around in the old school bus, we arrived at our designated motel just at nightfall.

We were all tired and hungry, having finished our bananas a long time ago. Although it was still early in the evening, there did not seem to be too many other patrons around and we were all dismayed to find the dining facilities closed. Lynn, who seemed to be the tour guide-in-charge, disappeared into the kitchen and was able to scrounge up something for our group to eat. No one was happy, yet Hugo and I tried our best to take everything in our stride.

The dissatisfaction grew when we checked into our assigned rooms. Upon opening our doors, we were overwhelmed by the gagging smell of permeated mustiness. The smell reminded me of my uncle's bone meal cattle-feed plant in South West Africa (later renamed Namibia) where I lived in my tender years. Chuck had informed us earlier in the day of the importance of good cross ventilation when building a house, to minimize the unpleasant effects of mildew that occurs quickly in the high humidity of Belize. Whoever built the motel had definitely not taken that precaution, and those members of the group who were accustomed to more sanitised living of North America, were not at all pleased with the situation. It was an ominous start to the tour and we still had two more days to go. Fortunately Hugo and my South African pioneering heritage stood us in good stead and we viewed everything as part of the experience in a tropical developing nation.

On day four we said goodbye to Lynn who remained in Corozal, and continued the tour with Andy and Judy as our guides. We flew back to Belize City and this time we landed at the small municipal airport with the runway right on the calm water's edge, where long legged herons, oblivious to the noise of small planes landing and taking off, stood patiently waiting for a catch. We boarded the bus with the addition of a friendly and knowledgeable Belizean tour guide who informed us that he had partied too much the night before. The bus squeezed through the narrow congested streets of the dilapidated-looking city before reaching the Western Highway, and headed west in the direction of the Guatemalan border.

Our first stop was at the Belize Zoo, twenty-one kilometres from the city. It is situated on twenty-nine acres in the middle of the bush, and houses only rescued indigenous animals.

Belize Zoo

An American woman founded the zoo in the early 1980s as a way of encouraging local Belizeans to appreciate their wild heritage.

We arrived to some consternation amongst the zoo staff as they had just noticed a Jabiru stork with an injured wing, and the expat vet from Corozal who was usually on call, was out of the country. Andy came to the rescue by volunteering Hugo's veterinary expertise. With only a roll of tape available, Hugo was able to tie up the wing but the long-term prognosis was not good. It crossed our minds that if we did retire in Belize, the zoo would be a worthwhile place where Hugo could volunteer his services.

Back on the bus, bumping over the potholes once more and water still dripping from the dysfunctional air conditioner, the dissatisfaction amongst the passengers started to take on a new life. The previous night in the musty-motel-without-dinner had primed people's discontent; some were wondering aloud why the tour's sponsors, Bill and Claire Grey had not yet made an appearance as they wanted to lodge their complaints about this tour not living up to their expectations. There were whisperings that Bill Grey might in actual fact be Andy and that Lynn who was no longer with us, was Claire.

We were a motley group form all walks of life. By all the complaints emanating from various parts of the bus, it seemed to us that we were the only couple seriously considering retirement relocation. Some couples were looking for a tax haven and one couple was looking for a place to continue their small aircraft manufacturing business away from the United States to escape the common hazard of being sued when negative consequences occur that involve one's product, whether one is responsible or not. Overall, most of the couples were non-committal.

Two other not-so-young single men did not hide the fact that they had come to Belize in search of women. At the end of the tour they were headed for the Princess Hotel and Casino in Belize City where all the action could be found. To my mind they must have been "Losers Back Home", a phrase borrowed from a friend in the American Foreign Service. One of them was seated in front of me on the bus, giving me a clear view of a comb-over that was held in place by some sort of product that had been applied not too recently and had left big globs of yellowed adhesive on his remaining strands of hair; all rather horribly fascinating to look at.

Next, we reached Belmopan, the smallest capital in the world with a population at that time of about 7,000. Andy encouraged the bus driver to hurry on, so we did not stop but made a quick trip around the Ring Road which circles the city, and then continued westward. Too preoccupied with making mental notes of the sights we passed, it did not occur to us to wonder why we had not stopped at any local supermarkets or other businesses except the open-air market and one hardware store in Corozal.

Our next stop was Spanish Lookout, the largest of many Mennonite communities scattered throughout Belize. They are hard working and productive communities and provide Belize with most of the staple food, poultry, dairy and building needs. The Mennonites literally feed Belize. We toured a site where prefabricated wooden houses of various sizes are built. Many people use them for temporary housing or as cottages, and can choose whether they want them placed at ground level or raised on wooden or concrete pylons. They seemed reasonably priced and delivery to anywhere in Belize could be provided.

It was early evening when we arrived in the town of San Ignacio; an older dusty Hispanic town with narrow streets situated against a hill. It is also the most westerly town closest to the Guatemala border. We checked into the decent-looking San Ignacio Hotel and happy to be in a hotel more up to standards we were used to. The rooms were attractive, fresh smelling and the bathrooms were done in colourful Mexican tiles. The hotel was proud of the fact that Queen Elizabeth II had stayed there on an official visit to Belize in 1994, and proudly displayed a framed local newspaper article that stated, *Di Queen shee eat rat* (The Queen she ate rat), referring to a gibnut which is a large rodent much favoured by the locals.

This was the hotel of choice in a town where there was not much to choose from. Again, there were loud mutterings of discontent from some in the group due to a mix-up in the hotel reservations; not everyone could be accommodated in the San Ignacio Hotel and some in the group had to be accommodated in a joint down the road. Hugo and I were relieved it did not include us. By this time Judy felt obliged, because of all the insinuations, to admit that Lynn and Andy were indeed Bill and Claire Grey. We were surprised at this revelation and some in the group felt strongly that they had been deceived and became quite vocal in their negative opinions of Andy, a few even referring to him as a reptile.

The next morning a few in the group flew on to Punta Gorda in Southern Belize. We forfeited that leg of the tour as we knew we would not even consider living that far from amenities. We liked the country-side around San Ignacio and thought it held potential for us. We finished our trip with an interesting guided tour of Xunantunich, a small but beautiful Maya ruin west of the town. The bus then took us back to Belize City from where we would fly to Vancouver the following day. We were dropped of at the Biltmore Hotel, a place that was adequate enough although she had seen better days. The two losers-back-home, now all decked out in fancy suits and carrying briefcases, were on their way to the fleshpots of Belize City.

***

Our interest in Belize as a retirement location was born out of two previous holidays spent on San Pedro. The first time was in 2000 with a small group of veterinarians of whom Hugo was a member that formed a casual association combining scuba diving and continuing education. We stayed in a charming hotel called Ramon's Village in the centre of San Pedro and right on the beach. We all dived in the mornings and the men held their meetings in the afternoons under the palm trees while we wives relaxed by the pool sipping Ramon's famous margaritas.

Our second visit was with our sons William and Hugo Jr., in 2002 when we rented a condo on the north side of the cay. We were enchanted by the casual and informal lifestyle of the island where there wasn't a need to dress up or even wear shoes. In the evenings we would walk along the beach to dine at any one of a small selection of restaurants situated close to the waterfront.

One evening we chose a recommended restaurant owned by a British expat couple. Judging by the décor, they had lingered in many different parts of the world. Before one ascended the steps into the restaurant, a footbath was conveniently placed in which to rinse the sand off one's feet. The owner acted as headwaiter and I was charmed to notice his bare feet protruding from underneath his long French-style white apron.

The next morning, while riding around town in a rented golf cart (the general mode of transport on San Pedro), we again saw the restaurant owner. He was walking briskly down the dusty street wearing neat colonial-style khaki shorts and a shirt with his briefcase tucked under his arm and still in his bare feet. This did seem like paradise. The turquoise ocean was warm and inviting, the barrier reef running parallel to the mainland guaranteed good diving, and for us the palm trees always added a special ambiance.

During this second visit we flew to the municipal airport where the price of a ticket is half of what it is to and from the international airport, even though the distance is the same. From there we rented a SUV and took a day trip to the south coast of the country. From Belmopan one turns south onto the Hummingbird Highway, which winds through the Maya Mountains towards the Caribbean coast. The lush tropical jungle is thick and spectacular and the sparse population seemed appealing to us.

It was quite incongruous to see orange groves lining the valleys and pineapple plantations growing on the mountain slopes where the jungle had been cleared, and large cohune palms in the background. We drove as far as the Garifuna town of Dangriga on the coast, before heading back the way we came.

On the way we stopped at the Belize Zoo. It is quite unique in that the habitats are kept as natural as possible and one has the feeling of walking through the jungle. There are photographs of the many celebrities who have visited the zoo and no doubt have contributed to the donations that keep the zoo operational.

We found the local Belizeans to be friendly and hospitable people with their ethnic and cultural diversity. We must have looked our age, as many Belizeans in casual conversation enquired if we were going to join the Qualified Retired Persons Program (QRP) and move to Belize. We did not pay much attention to the idea and returned to the reality of our everyday lives on Vancouver Island, coping with seasonal adjustment disorders during the grey, wet winter months and impatiently waiting for summers to arrive.

Chapter 2

Twenty years from now you will be more disappointed by the things you didn't do than by the ones you did do. So throw off the bowlines, sail away from the safe harbour. Catch the trade wins in your sails. Explore. Dream. Discover.

Mark Twain (1835-1910)

We never consciously decided to retire but toyed with the merits of Hugo hiring younger veterinarians to work in his practice while he continued to run the business side of things. He was after all, only sixty-three and never considered coming to a complete standstill. However, as the poet Robert Burns said, "The best laid schemes of mice and men often go awry"; retirement came knocking on our door. For a number of years a progressively troublesome back pain had plagued Hugo and the eventual diagnosis towards the end of 2002 revealed he needed back surgery. We knew there and then that he had to retire from practise, and our life, as we had known it for forty years was suddenly over.

After coming to terms with this seismic change, we felt an urgency to create something different and exciting, an adventure to stimulate our minds and bodies. Moreover, the longing for a place in the sun, a longing that never leaves the likes of us who were born under the South Africa sun, crept into our thought processes. We considered the possibilities of various tropical locations known to us. However, all things considered, Belize already familiar to us, seemed to present the least logistical difficulties. We decided to explore the possibility more in depth and signed up to attend the Belize Relocation Conference and Tour in January 2003.

One of the conference speakers of particular interest to us was a certain Mr. Gomez. His business, International Services Limited, undertakes to negotiate on his clients' behalf, the bureaucratic red tape, applications and registrations necessary to apply through the Tourism Board for QRP status. That included a one-time fee for which we would receive an ID card, exempting us from having to pay the required non-resident's monthly visa fee.

Once registered under the QRP program, we would receive the same considerations as residents do, and we were entitled to bring in duty free all our household goods, plus a vehicle, a boat and a plane. We did not have a plane and the only boat we had was a kayak. Another important incentive was that Belize is officially an English speaking country with a democratic government and unlike some of her neighbours, did not have any coups or insurrections in her history. We also liked the close proximity to North America so we could keep in easy contact with our children and our ties to Canada.

The most important factor for us in choosing Belize was that there would be no difficulty in taking our pets into the country, as the only requirement for them was the standard health and rabies certificates. One of the expat speakers suggested one should go to Belize for a trial period first in order to experience the two different seasons; the dry season from February to May and the wet season from May to February, with some variableness. That was good advice, but rather impractical for us as we could not see ourselves schlepping two cats and two dogs back and forth between Canada and Central America for a trial period. Therefore, we disregarded that advice and decided we would take the plunge and hope for the best.

After reading and gathering as much information as we could about living in the country, the decision to relocate to Belize was made. It all seemed like such an exciting adventure to undertake.

***

Two thousand and three turned out to be a busy year for us. In January Hugo's practice sold without too many complications. In May we went to Maui to celebrate Hugo Jr., and Alison's wedding. As her hometown was in Florida, we had not as yet met her family so coming together in Hawaii for a few weeks was a perfect occasion for the Miller and Lambrechts families to become acquainted. In June Hugo underwent surgery. Next, was to sell our home of twenty-one years. Before we even listed the house with a realtor, word of our impending move had spread through the community grapevine and we received a request from an acquaintance to view our house. Three days later we had a sale agreement.

As the new owners would only take possession in September, we had all summer with ample time to sort through and pack up our goods in Canada while Mr. Gomez set about submitting QRP applications in Belize. We could not believe how easily our plans were falling into place, and felt providence was smiling on us.

We were downsizing. We would be leading a much different and simpler lifestyle in the tropics and would not need most of the fancy household items we had collected over a span of nearly forty years. We set about sorting out what we would and would not ship to Belize. Apart from regular household items, we decided to ship all our large and smaller electrical appliances, as well as certain sturdy pieces of good furniture.

We held a garage sale for the remainder. However, this was not your ordinary garage sale. It was a by word-of-mouth-only invitation to friends and their friends. There were too many items to put price tags on and we had no idea what to charge. We solved that problem by putting out a large jar labelled "Retirement Fund". Our friends were free to donate whatever they felt their purchases were worth to them.

We put "for sale" labels on the furniture we were not keeping, set out the smaller items on display as much as possible in the appropriate locations: all the china, crystal, glasses and table linens on the dining room table; kitchen items in the kitchen; one bedroom was turned into a boutique of sorts where I set out our good winter clothes we would not be needing. I dressed up a display-window dummy in an imported

French skirt and sweater; set out my many pairs of Italian made shoes and friends were free to try on and make their choices in private.

There was a lot of bric-a-brac to put outside on our deck; Christmas decorations; tools; camping and fishing gear; emptied photo albums and much more. People came and went for a week. A few people, whom we did not really know, came back too often and a few early birds rang the doorbell even before we were up in the morning. We gave many items away, including boxes full of books. We had generous friends and at the end of the sale we were quite satisfied with the outcome, or more aptly, the income.

We surrendered some antique pieces of furniture, objects d'art, Hugo's hunting trophies and other African objects collected during our heyday in South Africa to an auction house. His trophy buffalo horns, which took a lot of effort by my sister to send from there to Canada, were the first to be sold. I wondered why someone would sport a trophy on their wall that they themselves had not actually been responsible for shooting.

Some people suggested it must have been difficult to part with so many of my possessions, but I surprised even myself by remaining quite detached from it all. A new adventure unencumbered by so much materialism lay ahead.

In early September the moving company arrived and the time had come to leave the home we had built, loved and enjoyed. Our acceptance papers had not yet been processed and we had to wait until we received the go ahead from Mr. Gomez as to when our container and we could depart for Belize.

***

To wait out the time, our friends Hermann and Tracy kindly offered us the use of their cottage on the lake. With a fax machine and a laptop, we comfortably settled into the cottage while waiting the call from Belize. In the mean time we read more books about the country and the dangers of travelling though Mexico by car. We had decided driving was the safest way to transport our two West Highland Terriers and two Burmese cats. We looked up pet friendly motels en route and planned our trip south, a journey Hugo estimated would take us about two weeks travelling at a comfortable daily rate.

We bought two 2'x3' collapsible wire dog crates; one to house the cats on the journey. Their litter box, which had a lid, could be zap-strapped into the crate doorway so as to block their exit and at the same time give them their toilet access, a sort of cat-en-suite if you will.

We relished the idea of seeing parts of the United States and Mexico on the way and anxious to depart before getting caught by inclement winter weather in the mid-United States. It was therefore with great relief in early November we were faxed all our necessary Belize entry documents, except for the QRP ID cards. After many fond farewells to friends and acquaintances, at last we were able to depart.

We left the cottage in great disarray; we had more than we could fit into our SUV. With the help of our friend John we repacked and had to leave some things behind. In our hurry to catch a ferry, we had no option but to leave our mess for Hermann and Tracy to clear up. We were packed to the hilt and everything had its place, fitted like a jigsaw puzzle.

The dog basket fitted on top of our overnight bags, level with the passenger seat headrest. That gave the dogs a sort of high window seat.

On our way

The cats seemed comfortable in their crate. With a new bright orange Nissan Xterra, a yellow trailer filled with luggage towed behind and a luminous green kayak on the roof, our psychedelic-looking ensemble was ready to go.

***

On November 15, 2003 we crossed the Canadian-United States border from British Columbia into Washington State. We were waved through with no questions asked. The pets had settled down better than expected and we anticipated a pleasant journey. Our first night's stop after crossing the United States border was Pendleton just inside Oregon, which we reached after dark. We realised we needed to arrive at our daily destinations earlier in the day, as it was stressful to unload the vehicle and pets, especially the cats after dark.

The daily routine was that we had to cut the zap straps off the litter box and quickly move the portable pet carrier into that doorway, so that the cats could not make a dash for freedom and escape. The following morning the cats had to be moved back into their travel crate, the litter box reattached and everything else packed back exactly as before in order for everything to fit. That process had to be repeated every day.

It was quite an undertaking. Mango was calm and easy to manoeuvre, but Guava panicked easily and would have a time of complaining every morning after we set off. The dogs were well behaved, except that Moraig eventually started growling at Lewis every time he bumped against her up high in their small basket. Interesting that the females of the species fussed and the males endured without complaint. Nevertheless, their general stoicism throughout two weeks of confined travelling and being in a different place every night gave us a new respect for their spirit of trust and endurance.

Thus, our interesting journey south through the United States continued without any mishaps. Our south-easterly route took us further through Idaho, Utah, Colorado, New Mexico and Texas and we didn't have any trouble finding clean and comfortable pet friendly motel accommodation. Every state we crossed seemed to have its own unique and beautiful landscape. When we reached Moab in southern Utah, we decided to spend two nights there in order to rest up, do laundry and have time to visit spectacular Arches National Park. Moab had a friendly atmosphere and our motel was comfortable and we found a good supermarket where we could buy take-away meals. After Moab we started to share the driving to make more miles per day with less individual fatigue. We stopped for lunch at the Great Divide in New Mexico and Hugo was sure he could see the bump.

We arrived in Albuquerque by late afternoon and felt more rested. With shared driving we could average about five to six hundred kilometres each day. Cheap motels are generally not in the best part of town, so eateries in those areas are not necessarily the choicest either. Near our motel in Albuquerque, looking for a place to eat, we found a diner that gave new meaning to fast food. With a cafeteria-style tray one had to wait in line for a server to spoon food, some of unrecognisable origin, onto one's plate. We took one look and walked out faster than we had come in. We decided on a restaurant close by that advertised Mexican food. Hugo thought that would be a good introduction to the cuisine that lay ahead. He enjoyed what we ordered and I thought he had to have been hungry to do so.

By now we could feel we were entering a warmer climate zone so we were happy to pack away our winter shoes in exchange for summer sandals. The excitement of a life in a tropical climate increased as we continued south. Our route was now less on secondary roads and more on interstate highways. We had never seen so many semi-trailer trucks on six-lane highways before.

I was quite surprised to realise that my husband was prepared to show some vulnerability by preferring that I drive the beltway around El Paso and other large cities where the truck traffic was almost bumper-to-bumper. I had always considered myself a better city driver, even though Hugo might not have acknowledged that. When we reached Fort Stockton, a proverbial one-horse town in Texas, Hugo noticed the new trailer tires were showing canvass. Fortunately, we were able to find a tire shop early the next morning and had trailer tires replaced with radials. We were now about half way to Belize.

On November 21 we arrived in Corpus Christi, Texas, on the east coast of the United States. It had taken us six days to travel from the Pacific North West to the Southern Atlantic. We were very thankful that we didn't have any mishaps along the way. We hoped for the same on the journey that lay ahead through Mexico.

In our readings we had come across both positive and negative reports of travelling through Mexico. Some were downright scary. We decided to spend two nights in Corpus Christi to rest and get organised for what lay ahead. The pets seemed happy to have a break from their daily confinement too. The motel was not too clean, but by this stage I had become less fussy. We did laundry, bathed the dogs and packed away all the remainder of our winter clothing. It felt good to be in summer mode in November.

Texas gave new meaning to "big". It catered to masses of people to the extent we had not experienced in Western Canada. Many signs were in Spanish. We had never been in a Wal-Mart before and this one was even a Super Store. We were quite astounded at the size; however, it gave us a good selection of cat litter to stock up on as we did not know where we would find any next. Then we found a supermarket, again "super-size". With so much food available, I thought no wonder so many of the population seemed super-sized as well.

For the most part we had been eating snack or take out foods. Since we were now in Texas cattle country, we craved a good steak dinner. Someone directed us to what appeared more like an "eating barn" with a revolving door. Folks lined up and were ushered in without delay to tables as soon as they became empty. There was no sociable lingering over a meal, only eat and go. The tables lacked place mats, making for quick and easy clean up. The waiters were quick too and everything moved like a conveyor belt.

After our stay in Corpus Christi we had a short journey to the border town of Brownsville. The motel we checked into was a well-known chain, but in this town it was bare bones. We noticed the closer we got to the Mexican border, the motel and restaurant standards were dropping considerably. Margarine instead of butter was standard fare at eating-places. By this time we learned if we wanted real butter with a meal, we had to take our own. I wondered if this was all designed to help me in the process of lowering my expectations for what lay ahead.

In Brownsville we had the Nissan serviced and were relieved that the radials on the trailer were still in good shape. The next day we went to Sanborn's; the recommended vehicle insurance agency and source of all information for travelling through Mexico. This was a vital step for anyone crossing the border. Ignorance or mistakes in Mexico could easily land one in unexpected trouble. Sanborn's provided us with maps and all the needed advice, including not to tell the Mexican border officials we were travelling through to Belize as they would turn us back to another border crossing hours away. Only the Mexican bureaucrats knew the reasons; we were to say that we were on an extended holiday through Mexico.

We had no trouble finding pet friendly motels in our travels, but we were a little concerned about Mexico. Sanborn's assured us that we would have no trouble as no one there minded pets. In my Internet search, the only pet friendly hotel that came up was a Howard Johnson in the city of Tampico. We also received word from Mr. Gomez's office in Belize City that our accommodation there was awaiting our arrival. We were excited at the prospect of nearing our destination and everything having been organised for us.

Chapter 3

Stop worrying about the potholes in the road and enjoy the journey.

Babs Hoffman (1922-2013)

At eight o'clock on the morning of November 25 we crossed the United States border at Brownsville into Matamoros, Mexico. After having read about all the problems one could encounter getting through the border, we were a little apprehensive; it was a mixture of excitement and fear. However, things went quite smoothly and the only irritation we experienced was a long wait to get our visitor visas and temporary car permit, all with the lie that we were going to tour the Yucatan Peninsula for five months. How else would we explain the cats and dogs, the kayak and all the contents of our trailer?

At the border the immigration officials were stone-faced; no one spoke English and there weren't any signs in English either. We thought that the Mexicans might have a reciprocal attitude, considering that there were so many of them in the United States and where most signs are in English and Spanish. According to our experience, it seems that unfriendliness is part of the universal immigration officials' job description. I could only think that perhaps the Mexicans did not like gringos.

(According to *Urban Dictionary*[3] the term "gringo" is used to denote foreigners, often from the United States. The word gringo is not necessarily a bad word. It is slang but is derogatory only in its use and context. From here on I will refer to all Caucasian foreigners in Mexico and Belize as gringos).

Matamoros was a glaring contrast to Brownsville. It was hard to fathom how one river could make such a difference between orderliness on the one side and chaos on the other. We had been forewarned that Mexican customs would be about thirty miles further into the country. Some books told of harassment, intimidation and even theft by officials. We had to be prepared to smooth our way with bribery, *mordida* (literally meaning "the bite") in the Latin America vernacular.

When we reached the Mexican customs, we were surprised to find there was no building as such, only a checkpoint with a group of scary looking men in uniforms carrying guns. They made Hugo unlock the trailer, while I wondered if we would have to unload and unpack everything. After a few tense moments when they opened some bags and found only books and clothing, they seemed to be satisfied and waved us on. We were relieved: so far so good.

We continued our journey, which would take us mostly on secondary roads closer to the Caribbean coastline and across the southern end of the Yucatan Peninsula and into Belize. This was a shorter route than travelling along the Pan American Highway, the latter being the recommended one. It did not take long to realise why. The roads were narrow and bumpy with no shoulder. People drove like maniacs and overtook on blind rises and corners. It was therefore not surprising to see frequent small shrines along the roadside, marking the spot where a loved one had sacrificed himself or been sacrificed by someone else to the speed demon.

One also had to beware of the "sleeping policemen" or *topes*, as those dangerous commercial-sized speed bumps are called. Hitting one of those at a high speed would render ones vehicle instantly un-roadworthy. The countryside was rather lovely and we were surprised at how many large cattle ranches were in the area. Huge billboards in the shape of an ox advertised the fact. We therefore anticipated a nice relaxing steak dinner in the hotel at the end of a long day.

Our first night's stop was the city of Tampico where we arrived during rush hour. The narrow one-way streets were bustling with people, and cruising taxis came to a halt wherever they were waved down, or beep-beeped at anyone they thought might be a potential customer. By now I knew Hugo did not like driving in such mayhem and he made his feelings known. This was our first real stressful situation in Mexico.

We drove in circles down the maze of one-way streets with people gawking at our strange looking vehicular ensemble until we eventually spied the pet friendly Howard Johnson hotel. There wasn't a designated hotel parking lot as there had been in the United States; therefore, making all the unloading and transferring the cats and all our other accoutrements rather nerve-racking. This time we stopped in front of the hotel on the narrow street. Hugo went to check in, only to be told that they did not allow pets, in spite of what we had read. Pleading our case and stopping short of accusing them of false advertising, the head honcho relented and allowed us in. The general staff were friendly in helping us unload our strange cargo, while passers by continued to gawk.

The vast hotel foyer and reception area looked quite fancy with marble-like tiled floors and huge pillars. There did not seem to be many other patrons. Even so, the room assigned to us was definitely not a reflection of the foyer. The single window was boarded up and the only ventilation was a small panel of louvers built into the bottom of the door. The air conditioner pumped musty smelling air into the room, reminiscent of the motel in Corozal. The bathroom was tiny with one small threadbare towel and a few dead cockroaches on the floor. We were grateful to have been given accommodation at all so we certainly were not going to complain.

Once settled in, Hugo took the dogs for a walk. There was a plaza on the corner of the block, but the little patch of lawn was chained off and our dogs were not used to using concrete facilities, so he returned to the room with mission unaccomplished. We went to the dining room to console ourselves with the anticipated steak dinner and perhaps a glass of wine, only to be told there was no *carne* (meat) on the menu that night. We eventually settled for some nachos and a beer. Judging by the lack of people in the dining room, the hotel definitely seemed empty. Our assigned room was most likely kept just for undesirables like us.

Just before bedtime, it was time to take the dogs for their last walk of the day. In desperation, Moraig did her jobs on the sidewalk, which Hugo, like a good citizen, picked up with the plastic baggies he always carried, but Lewis held on. As Hugo walked past the reception desk on his way back to the room, to his great embarrassment Lewis decided to stop and do his jobs right there in front of the reception desk with the management looking on. That must have justified to them why they did not want to accommodate us in the first place. Hugo had taken some Spanish lessons prior to leaving Canada, and he knew enough to make a sheepish apology. Back in the room, the vision of that incident somehow broke my pent-up tension and I could hardly stop laughing. I suggested to him that the embarrassment was good for his character, as it had given him an opportunity to practise some humility.

Needless to say we were up and out of the hotel very early the next morning and we looked forward to a better night in Veracruz. Again, we thought the scenery was lovely and noticed many cattle ranches. Our route took us on stretches of toll roads that were in good condition, but the secondary ones were riddled with potholes. We passed many small and rather impoverished villages that were guarded on either side by very large *topes*, an obvious sign of lack of traffic police. It was common to see disabled people and beggars sitting next to a *tope* where one was forced to slow down almost to a stop, giving an able-bodied *amigo* an opportunity to stand in the middle of the road and shake a tin in one's face if the window was open: we made sure to keep our windows closed.

***

All the bumping and shaking over the potholes was very unsettling for the pets and Guava especially complained. We were therefore pleased to reach the seaside city of Veracruz at about three o'clock in the afternoon. It seemed like a charming city compared to Tampico, with wide-open streets and some grand and opulent hotels along the foreshore. We drove around checking at all the hotels on our list. They were all adamant about not allowing any pets. Even the dubious looking places that seemed vacant refused us.

We drove around the city for three hours, searching in vain. By the time darkness descended, in desperation we accepted a villa at an expensive waterfront hotel that did not allow pets; with their proviso that we leave the pets in the vehicle overnight. We could park in front of the villa, where the night watchman would keep an eye on them. Leaving the animals in the car, we looked for dinner at an open-air restaurant across the street. We were again told that there was no *carne* on the menu. What was up with the Mexicans? They had so many cattle billboards but no meat on the tables. We settled for pan-fried snapper that appeared whole on our plates, causing us to fight our way through skin and all the bones in order to find the meat.

By this time the pets were restless and most likely could not understand why they were still confined in the car after dark. Hugo gave the watchman a good tip, and as soon as he disappeared around the corner, we quickly did the unloading routine and sneaked the pets into the villa. In all the subterfuge commotion, Guava made a dash for freedom and I was able to grab her just as she was about to jump off the back bumper. That nearly caused me heart failure. The villa was rather luxurious inside, but to avoid detection and eviction we had to keep the dogs from barking at unknown noises. Again we were up and away early the next morning.

***

This was our third day of travelling through Mexico and thankfully our last night in that country. By now we felt, contrary to what Sanborn's in Brownsville had told us, that Mexico was not a pet friendly place at all. We saw many pitifully neglected dogs when passing through the small villages and countryside. Villahermosa was our next destination and we anticipated the same accommodation difficulties as before. We were feeling quite stressed. So far we had not experienced any of the other travelling horror stories we had read about, but these accommodation difficulties were bad enough for us.

The further south we travelled, the more lush and tropical the countryside became and the roads were still as bad. The poverty of the small country villages was in stark contrast to the wealth in the large cities. In Villahermosa we encountered the hotel refusals once again. By this time Hugo was fed up and continued to drive on, determined that we would sleep in the Nissan at a petrol station somewhere along the way. We stopped to fill up at the first station beyond the city, only to be told that the next station was 390 kilometres away. We had read about the many dangers of driving after dark in Mexico. If a gringo was involved in an accident, it was straight to jail, guilty or not.

Hugo and I exchanged some heated words. Sanborn's had given us a triptych for Mexico that had lots of useful information in it. We found the name and address of a small hotel in Villahermosa that was owned by a Scottish expat. I suggested to Hugo that if he saw our West Highland Terriers, he was sure to give us accommodation. We therefore turned back and headed to the hotel, only to be told that the owner was out of town. We explained our predicament to the receptionist and she offered to help us. After making several phone calls, she found a hotel that agreed to take us, but only for one night. The Hotel Baez Carrizal was a tall building in an industrial area. Judging by its bright orange and cobalt blue exterior and even more colourful interior, it could easily have been designed by the Spanish architect Gaudi.

The staff could not have been friendlier. Many of them came out to greet us, helped to park the car and trailer in a secure area, and carried our strange cargo inside. We had a nice room with the best ambiance so far on the trip, and we were able to relax and enjoy that longed for steak dinner, even though it was slightly on the tough side and a little overdone. By this time we were grateful for small mercies.

After a restful night and with much lighter spirits, we left Villahermosa with the joyful anticipation of reaching the Belize border by afternoon. All things considered, Mexico had not been the best experience for us. We had found it rather frustrating not being able to speak the language, although Hugo's smattering of Spanish did help somewhat. Some distance after leaving Villahermosa, approaching a *tope* near a village, we saw a rope with red flags strung across the road. Our first thought was that bandits were ambushing us, until we got close enough to see that it was beggars attempting to stop us.

After three tense days of travelling through Mexico I did not feel too compassionate; when they saw my shaking fist they lowered their rope and we continued. As we travelled across the bottom of the Yucatan Peninsula, the road conditions seemed to become relatively better as we came closer to Chetumal, the Mexican border town. The traffic was less and the population became sparser. We also started noticing the traditional Maya stick-houses that seemed to be maintained with a better sense of pride than the previous poor village grass shacks. The vegetation was lush and colourful bougainvilleas seemed to thrive on their own.

Chapter 4

Travel and change of place impart vigour to the mind.

Seneca (54 BC-39 AD)

It was with great relief that we arrived at the Belize border at 3pm on Friday November 28. It felt good to be understood in English again. The Belize border formalities took about three hours. Hugo was a little taken aback when the Belize Agricultural Health Authority (BAHA) official questioned how he could be sure that Hugo was really a veterinarian and that our pet permits were legitimate. Since one's occupation is stated in one's passport, that question seemed rather bizarre.

Moneychangers do their business at the border crossings, so an official took Hugo to one to exchange our Mexican pesos into Belize dollars. Hugo gave the fellow a BZ$30 tip, but that was declined in no uncertain terms and he demanded BZ$100, to which Hugo complied. Just newly arrived, we were quite ignorant about their bribery and extortion habits, none of which had been mentioned in any of our readings about moving to Belize.

At the border are "runners" who go to the next point to get one the necessary car insurance, a dubious system to say the least. During all these formalities I stayed in car with the pets. The insurance runner returned to our car and asked me for a tip. I had no Belize money on me and told the runner to wait for Hugo in the immigration building. As it was getting late, it seemed he did not want to wait and left on the last bus. Unbeknownst to me, Hugo had already given him a substantial tip. This was the second attempted scam; we were starting to notice.

With darkness upon us, and all formalities cleared, we were free to drive on and into Belize. We were exhilarated. After having travelled for two weeks and covered 7122 kilometres, we looked forward to a relaxing night at the end of our journey. Judy, our guide from the Retirement Conference and Tour with whom we kept in touch, had booked us into a popular expat "watering hole", Nestor's, in Corozal. There would be no pet difficulties this night.

A short distance into Belize, we were stopped at an unexpected checkpoint. The official came to my window and told me I had just broken the law. When I asked, "How?" he informed me that I was not wearing a seatbelt. In all our excitement and the darkness, we had not noticed any sign indicating such. He ordered Hugo into the checkpoint hut and told him that as the driver, he was responsible for his passenger and verbally fined him $30. When Hugo handed the money to him, he said that that had to be thirty US dollars. Hugo thought that was rather odd but who were we to query anything at this juncture.

We felt it would have been more courteous to have given us a warning, since we were obvious newcomers and not yet familiar with the laws. We shrugged it off as part of the DNA inherited by some Belizeans from their British buccaneer forebears.

At last we arrived at Nestor's in Corozal. At first glance we could tell that it was a low budget establishment, but by now we did not care; nothing could be worse than our Tampico experience. After settling the pets into our tiny room where two of us could not squeeze past each other at the same time between the bed and the wall, we went into the pub-cum-dining room for a meal and a local beer, known as Belikin.

We were hungry and tired. We recognised many of the characters in the pub; they looked similar to the faces we had seen in our earlier days in outposts of colonial Africa. Belize, formerly know as British Honduras before independence from the United Kingdom in 1981, had her fair share of expats. People were friendly and many struck up conversations with us. Some were only “wintering” in Belize while others lived there year round. We had a good meal and went to bed relaxed for the first time in days.

The next morning we were up early and on our way to Belize City to meet Mr. Gomez at his office in the downtown area. He would guide us to a pet friendly hotel he had booked us into, where we would stay for a week until the rental house he had also arranged would become available. Mr. Gomez’s services had been invaluable.

***

Belize City, with a population of about 70,000, is a city like no other we had seen before. Chaotic was the first impression that came to mind. The streets were more like a narrow maze, lined with many ramshackle buildings knocked together with wooden planks and sheets of corrugated iron. Living abodes and businesses were all mixed together. Dispersed amongst them were dilapidated colonial buildings that had been architectural beauties once upon a time. Here and there we crossed over some of the many dirty canals that stagnate throughout the city and connect to either the Belize River or the ocean. After driving in circles and being gawked at, and we gawking back, we eventually found Mr. Gomez’s office.

The Bakadeer Inn

The small Bakadeer Inn where we had accommodation waiting, looked unimpressive from the narrow street, where we also had to find parking, but the inside looked clean and inviting.

The floors were tiled and the rooms were comfortable with ceiling fans. They looked out onto a fairly narrow, but bright courtyard enclosed on the opposite side by a high wall providing privacy from the close proximity of the adjacent building. Next to the Inn on the other side was a carpentry workshop that we fortunately could not see, but through our small bathroom window came the noises of sawing, hammering and loud music-not-to-our taste till late into the evening. We remained good-humoured, as this was all part of the experience.

The staff at the Inn was friendly and welcoming. It seemed like a family affair and we got to know them all by their first names. Gilda and her son Michael managed the front desk. With what little I knew at the time, I thought Michael was Creole. Gladys, the housekeeper was from Guatemala. I had not done laundry since Corpus Christi in Texas, and appreciated the free use of the hotel laundry facilities. Tulio and his brother rotated night watchman duties.

After a few days we got to meet the proprietor Mr. McField, who was equally friendly and made us feel welcome. At one point he told us an important fact to remember in Belize was not so much "what" one knew but "who" one knew. With time we came to realise what a truism that was. Whether it has positive or negative consequences for individuals depends on which side of the political divide they stand at that particular time.

Having unpacked our vehicle, the trailer and the kayak needed to go to Mr. Gomez's house for safekeeping. The pets no doubt sensed us feeling more relaxed, as their mood followed ours. By now the dogs had grown accustomed to doing their jobs without a lawn. Hugo took them for regular walks down the dirty narrow unpaved streets with smelly narrow open gutters in place of sidewalks. I could always tell whereabouts he was walking, by the proximity of all the other dogs barking in the area. We had never seen or heard so many dogs as here, all of mixed parentage and in various degrees of neglect. Every property had one dog or more. Many abodes were unfinished with little yard space, so some dogs would be barking from the floor of the potential second level.

We soon found out that Belizeans are afraid of other people's dogs. Theirs are kept mostly for security purposes rather than as pets. Just about everyone stopped, even people in cars, to ask what kind of dogs ours were and whether we were going to have *poppies* (puppies). Everyone wanted a *poppy*. Even though our dogs were neutered, after the neglect we had seen in so short a time, I would never have entertained such a thought.

The Inn did not have dining facilities, so for our first dinner in Belize City (eager to try typical Belizean fare) we were recommended to try Macy's on Bishop Street. It was not a restaurant as we were used to in North America. Instead it was one of the many "holes in the walls" of narrow streets, with simple wooden plastic-covered tables and chairs. It looked clean and the proprietors were friendly. We enjoyed a meal of fried lobster (available in season and many times out of) and a choice of rice and beans or beans and rice. Yes, there was a difference: the first one is when the two are cooked together and the second is when the rice is cooked alone and a bean stew served along side. We chose the rice and beans, as that was more traditional.

The following day was Sunday, and we decided to attend the Queen Street Baptist Church, which we had noticed near Mr. Gomez's office when we arrived in the city. It is an historic wooden church founded in 1850 by a British missionary, Alexander Henderson. We felt a bit strange being the only gringos in a mostly Creole congregation. Pastor Lloyd Stanford and his wife Nancy, who had served at that church since 1973, and who still played an important role in the Baptist and missionary community of Belize, warmly welcomed us. We found the service and music uplifting and noted with interest that everyone in the congregation still adhered to the by-gone tradition of attending church dressed in their Sunday best. That was quite a contrast to the affluent society from whence we had just arrived, where casual was the order of the day.

By now we longed for more familiar food and it was suggested we try the Chateau Caribbean. This hotel was situated on North Front Street opposite the ocean wall, and must have been quite stately in colonial times, but her grandeur was by now quite faded. The dining room was located upstairs on the first floor with a spectacular view of the Caribbean Ocean, slightly marred by the obvious polluted water. It was evident that time had taken its toll on the Chateau as now the wooden dining room floor had a slope and the carpet had seen more traffic than it could bear. The wallpaper was faded and stained and paint peeled off window frames. Nevertheless, we found it all charming and enjoyed slightly more familiar food with service and table settings reminiscent of a more genteel past, in spite of the little bun with of a pat of margarine, which they assured us was butter.

We decided to drive through the tidier residential areas and familiarise ourselves with the city. Princess Margaret Drive and adjacent areas were where the more affluent citizens lived. There were other parts of the city that looked pretty scary and we were told to stay away from Southside in particular. We decided to be cautious and always kept our car doors locked.

Many of the businesses and homes looked dilapidated from the outside and from the glimpses one could get, many homes seemed the same inside. Mr. Gomez's office was situated in the same street as the Embassy of the United States, a fairly grand-looking old building. His office looked smart inside and outside, but the floor above him had tenants whose washing was left hanging out to dry on the balcony. We didn't see any shopping malls and most of the shops, regardless of what they were named, had the same look as the old general stores of yesteryear. We had not noticed any supermarkets, but we took note of a parked lorry loaded with fruit and other fresh produce.

On Monday we had business to attend to. The first order of the day was to organise bank accounts. We had been advised to open and deposit money in a US dollar account before our arrival in the country. (Belizeans are not allowed to have US dollar accounts, but with the right connections, I am sure many do). Now, we had to open a Belize dollar account for general use, into which we could transfer money from the offshore US account. After that we met all the staff at Mr. Gomez's office, where we were welcomed as family.

Sandra, who had been responsible for handling our QRP file, would be taking us to see a rental house. I wondered what we would have to settle for. I did not expect American-Modern, but after seeing so many dilapidated houses in the city the previous day, I was apprehensive about what sort of house it would be. It was our intention to stay in Belize

City until we had found where we wanted to buy our own little patch of jungle.

Sandra directed us down South Front Street and it had its share of potholes. There weren't any shacks on this wide street, but most of the houses were colonial has-beens. We stopped at a more modern peach coloured, double storied, flat-roofed concrete house with a high wall surrounding a narrow yard. The sea wall and ocean were on the opposite side of the street. From there we had a good view of the comings and goings in the Belize harbour and the pelicans and other tropical birds flying around.

We were offered the ground floor of the house, which had tiled floors throughout, a nice looking kitchen with a tiled countertop and two bedrooms with bathrooms. The layout of the house and bathrooms was quite strange; the bathtubs-cum-showers were tiled rectangular built-in concrete containers, more reminiscent of livestock drinking troughs. Even so, the house was beyond my expectations and I felt greatly relieved. The high wall offered security for the dogs and us. On a windy day the sea spray would blow through the wrought iron front gate and into the bedroom window. The location of our rental house was perfect.

The upstairs of the house had its own outside entrance and was occupied by the landlord's housekeeper, Miss Pamela. To add Miss or Mister before any adult's first name is always a sign of respect in Belize, and we would be respectful. I was allowed the use of the washer and dryer upstairs until our container arrived with our own utilities. The landlord, who owned the local cable television station, was a friend of Mr. Gomez. He lived next door and we could have the use of a parking bay where his business vehicles were parked at night, with the benefit of their watchman; all that at BZ$800 a month (US$400). We were more than satisfied. That evening we celebrated with a Belize-style lobster dinner at the Chateau Caribbean.

We were to stay in The Bakadeer Inn until the end of the week when our rental house would be ready for us. We were informed that our container would be arriving around mid-December, just a few weeks away. That was good timing, as it meant that we would not be in an empty rental house for too long and would need to purchase only a few basic items to see us through.

During this week we received a visit from Ernesto, a friendly hyperactive customs broker who brought with him a man from Belize customs in order to clear our motor vehicle. Hugo was informed that his motorcycle had arrived, but he would have to pay duty on the latter. It was regarded as a vehicle and we were only allowed one duty free under the QRP agreement. We wondered if our two mountain bikes packed into the container would also be regarded as vehicles.

As we would only be unpacking necessities from our container, we decided to rent a mini storage in Ladyville, a new development on the northern outskirts of Belize City. The owner was an expat American, who had taken up with a young Hispanic woman and lived in a flat close by. I wondered if he too was a loser-back-home. A security fence surrounded the mini storage with a few pitbull-type guard dogs on the inside. If one wanted to gain entrance, one had to first locate the owner. We could now move our kayak and trailer from the Gomez residence to the mini storage.

Charles, a young Creole employee of Mr. Gomez, who was appointed as our "Man Friday" accompanied us. Two concrete strips across the lawn formed the driveway next to the Gomez house. It had rained heavily the night before. Hugo had to turn off the strips in order to hitch up the trailer. To our amazement, our SUV got stuck in the mud, camouflaged by the grass. It was impossible to straighten the wheels. After much futile revving and mud splattering everywhere, Charles went to look for help and returned with someone driving a tractor that was able to pull us out. That day we learned that Belizean mud gave true meaning to the word gumboots. With time we would come to learn how hazardous mud in the rainy season could be. At the mini storage we encountered another Belizean hazard: biting midges.

The week we remained at the Inn we had most of our meals at Macy's were the fresh fish fillets with rice and beans were tasty, or at the Chateau where the lobster was inexpensive, for obvious reasons.

Chapter 5

The first condition of understanding a foreign country is to smell it.

Rudyard Kipling (1865-1936)

As the rental house had no appliances and was not wired for my arriving electric stove, we purchased a small gas stove and a small refrigerator to be delivered to the house the day that we moved in.

We purchased a round glass-topped table and four chairs, which we knew we would use on a stoop when we eventually built our own house. We also acquired a cheap foam mattress to sleep on until our bed arrived. In the trailer I had packed a few basic kitchen utensils and some bed linen: we would be comfortable enough until our container arrived.

Through our landlord's cable company Hugo was able to connect his laptop computer. In order to have a telephone connection, as non-Belizeans we were required to put down a two thousand dollar deposit; the reason being that people absconded without paying their telephone bills, or so they said. Our shipping agent friend Ernesto purchased and sold us a cell phone, which enabled us to overcome that problem.

By now we had found a supermarket close by, although not quite North American standard but close enough. To inaugurate our new stove I decided to cook a beef oven roast. Even though the beef roasted for the prescribed time, and then some, it remained tough and retained a lingering bloody taste. We decided that we would stay away from Belize beef and stick to chicken, produced by the Mennonite community and the mainstay of most of the local diet. We were informed later that it was common practise to freeze beef immediately after slaughtering without first hanging it; hence the lingering taste. That habit gave the merchant increased profit as the frozen moisture added weight to the meat. Many business class Belizeans avoided local beef for the same reasons.

Our rental House

The first evening in the house, while sitting at our new table just off the kitchen area, from the corner of my eye I noticed a furtive movement across the kitchen floor, and then more and more. On closer investigation, I saw large cockroaches literally coming out of the woodwork. I knew they occurred in the tropics, and for good measure I bought a can of "Roach Raid" on our first shopping expedition. However, I was not really prepared to have them in my own house; there were limits to this adventure. I grabbed the roach spray and went to war. That night

I killed twelve. I acquired information on cockroach biology via the Internet (one has to know one's enemy) and learned that they like dark and damp places.

Well, there were many dark places in the wooden kitchen cupboards and lots of dark and damp areas under the concrete slabs surrounding the house. Henceforth I sprayed around the kitchen cupboards and around the outside perimeter of the house and doors every evening. Hugo thought the fumes would choke him to death. I noticed less and less roaches as the days went by, but I had to keep up the defence. I did not want to put any of my groceries or crockery in the lower dark kitchen cupboards where the roaches could leave unseen footprints. We bought a large open chrome shelving unit that had no concealed areas and where everything could be stored within sight.

One morning we awoke to find a small neat pile of what looked like coarsely ground white pepper on the kitchen counter. We were quite puzzled but logic told us that it must have come from above, although we couldn't see any little holes in the ceiling. Perhaps termites? When using the washing machine upstairs in Miss Pamela's apartment, I had noticed that the floor was made of wood, but there was a layer of gypsum board underneath the wood, which served as our ceiling. I took the evidence to Mr. Henry, the gardener, for confirmation. He informed me that those were indeed "termite seeds" and he nodded when I asked if he actually meant termite poop?

With this confirmation we had to do something to combat this added invasion. We decided to pin a sheet of plastic to the ceiling. The following morning the evidence was there; a neat little pile of "termite seeds" on the sheet. In the days to come we found more little piles around the doorframes and at the edges of the louvered cupboard doors, anywhere where there was wood. As fast as I swept them away every morning, they would appear again the next day. By now we had found out that these little termites, known in Belize as woodlice, do their own good housekeeping by cleaning out their abodes every night; they make tiny exit holes in the wood through which they push out their "seeds". One could almost characterise that as being woodlice "long drops". These termites slowly but surely devour any untreated wood structures in the country.

Having become tired of the daily sweeping, we systematically pasted duct tape over any frames that showed any evidence of housing woodlice.

As the house had very little garden area outside, Hugo took the dogs for frequent daily walks along South Front Street. He became acquainted with most of the tenants who occupied the old colonial homes, and everyone got to know the dogs that remained a novelty to the locals. One morning when we woke up, we noticed tiny grey blobs the size of pinheads on the wall above the bed. They were actually moving. On closer examination we noticed that they were tiny soft-bodied ticks. Lewis had obviously picked them up while they were in their invisible early developmental stage, while lifting his leg against a tuft of grass somewhere. Over the next few days they multiplied greatly and were crawling all over the walls and windowsills in the bedroom. We were kept busy getting rid of them. After that experience we tried to keep Lewis away from any areas where the local dogs where likely to have left their mark.

Not long after occupying the rental house, our container of personal effects arrived. The agents and brokers who were involved in clearing our goods were friendly and helpful. One of them was a British expat who was married to a Belizean, so he knew his way around the local customs and expectations. Then, there was an unexpected delay in releasing our container. In it Hugo had a safe with the remainder of his gun collection, which caused great consternation in the customs department.

When we attended the Retirement Conference in January, Hugo had been assured that there was no problem in getting firearms registered in Belize. I don't think anyone considered there to be more than two. Hugo felt confident now that they had approved his gun safe and put it in "Queen's Custody". There it would remain in safekeeping until Hugo had completed a separate importation application, which would have to be approved by the police department.

In due time the container, minus the gun safe, was released and brought by the broker to the mini storage where the customs officials did their inspection. They looked through our required detailed list of contents and scratched around in various boxes. They were not too happy that we had brought wooden furniture (as Belize has her own hardwood) but let the matter go when we told them we had not read anywhere that our personal household effects did not include our furniture.

When they had finished their inspection, Hugo had to drive them back to their offices. On the way they asked him for some money. When he looked rather puzzled, they informed him that it was customary to give the officials money for lunch after an inspection. They went away with a hundred dollar bill, known locally as a "blue-boy," between the three of them. We set about unpacking only the things that we would need while staying in the rental house in Belize City.

***

Belize itself is a true melting pot, and it is not always easy to identify everyone's ethnicity. The Maya are the original inhabitants in much of Central America and make up about 10 percent of the Belize population of approximately 335,000 (2012). They have maintained their languages of Maya Mopan and Q'eqchi, and are generally easily recognizable by their distinct features. The women wear mostly colourful dresses of a similar simple style.

Creoles are a fusion of British Baymen and Africans slaves. They were brought to Belize by the British from Jamaica and Bermuda as labour to cut logwood and mahogany in the mid-seventeenth century. They used to be the majority, but to their dismay they are now reduced to 25 percent of the population, having been displaced by Mestizos, a fusion of Central American Indians and Spanish who make up 48 percent of the population.

The Garifuna, who comprise six percent of the population, are descendants of shipwrecked African slaves, integrated with native Carib Indians who originally lived along the Orinoco River and the Lesser Antilles. They too have developed their own language and their own particular culture.

The Mennonites make up about four percent of the population. They are of German origin and speak a language called *Plautdietsch*, an old form of lower German. Starting in the early twentieth century, many left the Canadian prairie province of Manitoba and other global regions to settle in Mexico, from where they moved to Belize in 1958 and further south as far as Uruguay. Their primary reasons for moving were to prevent their children from being contaminated by a public education system and they also wanted to maintain their own social and religious distinctiveness. They have succeeded in holding on to their values and remain hard working, independent and self-sufficient communities without outside interference wherever they have chosen to settle.

Others in the Belize melting pot comprise of Indians (from India) Chinese or Taiwanese who are mostly shopkeepers; Lebanese descendants who generally form part of the political and business community; and finally Europeans and expats of other diverse origins, some who choose not to disclose too much about their backgrounds.

***

Belize City is situated on a swampy little spit of land jutting into the Caribbean Sea, so it is surrounded on three sides by water and the Belize River runs through the middle. In the mid-seventeenth century, British buccaneers used the coastal area as a hideout from where they could prey on Spanish ships. Apparently the Spanish galleons were too big to get through the barrier reef, which runs down the 386- kilometre coastline.

When the British Baymen and African slaves arrived, they settled around the mouth of the Belize River, to where they could float down the logwood and mahogany, and then ship further to Mother England. The Battle of St. George's Cay is still an important celebration in Belize, where on September 10, 1798, Baymen and Black slaves defeated a Spanish force which had for some time been attempting to capture and claim the land for Spain. After their defeat, the Spanish left for good and the country became known as British Honduras.

The name "Belize" was born on September 21, 1981, when the country gained independence but remained within the British Commonwealth. Creoles for the most part have English surnames; evidence of the country's colonial past. They developed their own patois known as Kriol English, spoken and understood by all Belizeans but largely unrecognisable to some of us who only speak proper English. Their skin tone ranges from dark to very pale, some even with blue eyes. A small minority of Creoles, who have traditionally held high positions in business and politics, are referred to as "Royal Creoles". The rest remain rather poor.

As Belize City is at sea level, there is always the threat of flooding, especially when a high tide coincides with heavy rain.

Fish market on a canal

Then the dirty canals and open drains, which are more like sewers, intermingle with the rain and the polluted seawater to create a toxic mix. A few years after we had moved away from the city, just such an unpredicted storm system caused most of the downtown area to be flooded by almost four feet of this mix. Children were swimming in the streets even though there were warnings about health risks. Most street level businesses were closed and it took about a week for everything to be cleaned up and establishments to reopen. After it rains, the city has a pervasive sour odour that hangs in the humid tropical air.

The city areas are generally referred to as Northside or Southside, the latter being the poorest and more criminal part of town and newcomers are advised to stay away. Because of the low elevation in this area, in flood season residents have to walk over "London bridges", the name given to old wooden pallets that serve as elevated walkways between homes and the streets. In the early 1980s the Canadian International Development Agency (CIDA) donated funds to build a sewer system in

Belize City[4]. According to a 2008 Biodiversity Reporting Award[5], twenty years later only 30 percent of homes were connected to the system.

The public health laws still allow a common practise of the past, where poorer residents can dispose of the contents of their "night-soil" buckets into the sea or waterways, provided they do so by five o'clock each morning. The report also stated that only 16 percent of homes throughout the entire country are connected to a sewer system, and 42 percent are connected to a septic tank.

In her past history, Belize City has endured fires, disease and deadly hurricanes. In October 1961 Hurricane Hattie almost wiped the city off the face of the map. After that disaster, the powers that were decided to build a new capital in a less vulnerable part of the country. Like Brasilia, the capital of Brazil founded in 1960, Belmopan was founded in 1970, fifty miles to the west of Belize City. However, unlike Brasilia, it never developed into a beautiful capital, even though it is euphemistically called "The Garden City".

No one wanted to move from Belize City and it was not until the Americans built their new embassy in Belmopan in 2005, that the other embassies followed suit. Belize City still remains the centre for many business and cultural activities. It has its own unique style and energy, amid a cultural blend that Creole Belizeans like to regard as more Caribbean than Central American. It is the largest city in the country and many government departments still remain there. Despite the crime, violence and poverty, residents have affection for their ramshackle city, which they fondly regard as their "Jewel by the Caribbean Sea", that phrase included in the national anthem. According to a CNN poll, tourists placed Belize City tenth on a list of "World's 10 Most Hated Cities".[6]

***

We were finding our way around. I was impressed at how quickly Hugo got to know the lay of the land. I found the thought of having to drive through that maze of mostly narrow one-way streets where one's wheels could easily fall into some deep and narrow cemented gutters, quite daunting. Driving in the city is not recommended for those who are impatient or prone to road rage. One has to drive very defensively, as one can never be sure that road rules would be adhered to.

Bicycles weave their way in and out of traffic in whichever direction they choose. Vehicles do not necessarily give one the right of way, or bother to stop at stop signs. We soon found out that the same rules do not apply to everyone. South Front Street was a one-way street, the arrow pointing away from centre town. Everyone on a bicycle rode in the opposite direction, towards centre town. Hugo followed suit on his bicycle one day, just to be stopped by a "tourist policeman" who made him turn around. His protests that others were doing the same were ignored.

There are three stoplights in the entire city. One of them is positioned just above the white line, so if one's vehicle is first in line, the only way to know when the light turns green is to rely on the vehicles behind blowing their hooters to let one know it is time to proceed.

Our rental house on South Front Street was within walking distance of the downtown area and we enjoyed exploring around the various shops to see what was available. The storefronts were narrow, but once inside they seemed to stretch back for miles. One store, "Hofius", owned by expat Brits, had most of what one needed in the line of small household and hardware items. Our first purchase was a small metal fan, as our rental house did not have ceiling fans.

As our arrival in Belize was in the so-called winter months, we had not yet experienced the hottest season. Chinese, Taiwanese or East Indians owned most of the other downtown stores. Much of their merchandise was not really acceptable for the North American market, but there were things available that we had not seen since the fifties, like plastic coated curtain wire. That was the perfect item we needed when we put up temporary window coverings in the rental house.

Eventually we discovered the two supermarkets that were closer to what gringos are accustomed to. However, it was not "one stop" shopping. One carried some items that the other did not and one never knew who had what. Items that we favoured were naturally imported from North America or the United Kingdom and they came at a price. There were no grocery items labelled as "Choice Grade" and it was humbling to realise that the "No Name" brand of tinned food I did not deign to buy in Canada, was now my brand of choice.

I was relieved to find real butter available amongst all the unknown brands of margarine and lard. Of the few brands of butter available, we liked the one imported from New Zealand the best. Whipping cream from the United States was a scarce commodity and I stocked up when it was available. Being ultra-pasteurised it generally lasted well beyond its "Best Before" date. The cashiers would sometimes ask me what I did with cream. It was definitely not part of the local cuisine.

One could tell when a new shipment of groceries was due as many items had become unavailable and many shelves were empty. I learned if an item was available, it might not necessarily be so in the future, so we had to stock up while the going was good. So far we were exceeding the US$450 monthly cost of living mentioned in the Retirement Guide by Bill and Claire Gray, even though we were only spending money on necessities. There was nothing to encourage "impulse" buying anyway.

We had not found an open market where we could buy fresh produce; what was available in the supermarket was often wilted. Therefore, we relied on the vendor selling fresh fruit out of his lorry who regularly parked at the same spot next to the road. As warned, we could not find any good coffee in Belize, so we started the day with freshly squeezed orange juice instead.

Belize has wonderful, inexpensive juice oranges: juice from two oranges filled one glass. The oranges look ugly from the outside and would not have been visually acceptable for the fussy North American market; they're used for the juice industry instead. Before consuming any fresh produce, I adopted a habit from my grandfather, of first soaking them (and other fresh produce) for a short time in a solution of potassium permanganate, which killed any potentially harmful germs and especially intestinal parasites.

Eventually we found the poultry shop run by the Mennonites, where chicken was fresh, inexpensive and tasty. That became our main source of protein. When we bought beef, we relied on minced steak. Although Belize water is supposed to be potable, most people drink purified water delivered in five-gallon containers twice a week. We were never sure exactly how purified the container water was, but it sounded better. It took me a long time before I even brushed my teeth with tap water.

Apart from locally made rum and beer, imported liquor in Belize is expensive. We were told about 60 percent of government revenue comes from import duties. Through all our explorations, we found that most of one's basic needs in the line of hardware, appliances and building materials were available, although not of the best quality. One could never be sure if the appliances were brand new or "lemons" from North America. We had read that small appliances were most likely "customer returns" from the United States. We experienced that truth when we bought a bread machine that only lasted for six months. For more personal items and clothing, I would have to wait for trips back to Canada or America.

***

Because of the shallowness of the harbour, cruise ships that visit Belize have to anchor a few miles off shore. The cruise ships create entrepreneurial opportunities for locals; from operating water taxis to being tour guides who take visitors on various excursions into the interior jungle, cave tubing, zip lining and visits to Maya ruins. Many tourists choose to explore the city, although there is not much that one can really regard as tourist attractions.

Even though most Belizeans are friendly, there are many who are past masters at the art of hustling and scamming. As this can be a nuisance, one can see "tourist police" walking the streets on days the cruise ships are lying at anchor, ostensibly to make sure tourists are not harassed or molested.

We always received a friendly greeting from the locals. On one particular day, while walking downtown, a tall, well-dressed Creole man approached us and enquired how we were and where we came from. Upon hearing that we were newly arrived retirees from Canada, he proceeded to tell us with what high regard he held Canada, and especially past Prime Minister Pierre Trudeau who he considered to have been a good friend to Belize; notably that he had provided financial aid to cover up the open drains. Our sense of smell told us that the job had yet to be completed. Either the money had been insufficient to complete the project or some of it had found its way into politicians' pockets.

One of the bridges over the Belize River is named the Belcan Bridge. The "Can" signifying where the money came from. There is also a Belchina

Bridge, another donor identification. After the hail-fellow-well-met conversation, we indicated that we had to move on. The man suddenly asked us if we had a dollar or two for him. We were taken by complete surprise as he did not appear to be wanting, and having by now been filled with a sense of well being towards the man, we complied.

We soon realised that we had been well and truly hustled: after that we did not give out any money. If anyone asked, we offered to buy them food. Those who were genuinely hungry accepted the offer and those who became angry were the hustlers. Sadly, sometimes grey-looking little children would come by our vehicle and beg for a *dalla* (dollar). One never knew whether they were beggars-in-training or if a parent was using them for that purpose. On occasion someone else would stop us with the by-now-familiar hail-fellow-well-met tactic but I learned to ask early if they were going to ask us for money. That would usually put them off stroke and they would move on.

One day when I was standing just inside the security gate of our yard, some cleanly sucked chicken bones came flying over the wall and landed at my feet. My first reaction was one of indignation. I looked out over the gate, to see a tall Creole man walking on. With indignation still in my voice, I asked what he did that for? He turned to say that he was giving the bones to our dogs and sounded quite surprised that I was not pleased. I immediately realised that in his culture he was being kind, so I thanked him. One has to see the neglected and pitiful state of most dogs in Belize to realise how giving cleanly sucked bones to dogs makes sense to a local.

Nowhere else had I seen as many dogs wandering around in similar states of obvious neglect and abandonment. They were emaciated, most likely due to a diet of bare chicken bones and with mega parasitic infestations. Apart from that, they were suffering from every imaginable chronic skin condition common in hot and humid climates. Many dogs had bleeding skins from the consequences of untreated flea allergy dermatitis, mange and staphylococcus infections, to having almost no hair left at all. One could see by their demeanour how racked with pain they were. We came to recognise an overall greyish pallor they acquired when they were close to death.

There are of course Belizeans who care for their dogs and keep them off the streets, but they are in the minority. We were told that tourists from the cruise ships complained about the sorry state of the many street dogs they saw around the city, so the city council became more diligent in their dog control. Their method of choice is to put out meat laced with strychnine poison on the streets at night, a very cruel and painful death. Neutering animals is not part of the Belize mindset, let alone when people are poor, so there remains the continual increase in the unwanted dog population. One sees very few cats, because they are generally disliked and the many stray dogs keep their population under control.

***

Just before our first Christmas in Belize, we were pleased to receive a wedding invitation from one of Mr. Gomez's secretaries. The church was conveniently close to our house and the service was to start at four o'clock. We arrived fifteen minutes ahead of time, worried we would be late, only to find we were the first guests to arrive. The groom arrived at four o'clock and the bride eventually arrived at four thirty. That was our first experience of what is generally referred to as "Belize time". The reception was held in a school gymnasium and definitely a different cultural experience for us.

Due to her religious convictions, the bride's mother would not allow any alcohol or dancing at the reception. The head table seated only the bridal couple and their attendants, while the bride's parents acted as door attendants. We did not know if the groom had parents; we never saw them. Two large stereo speakers blasted out music too loudly to allow for comfortable conversation with fellow table guests. Just as well, as we had to fight the "fry" chicken and rice and beans with a plastic fork that was served in a take out styrofoam container. It was not long before our ankles were attacked by the biting midges and we had an excuse to make a polite, but early exit.

Christmas approached, but it could just as well have been July. We welcomed the absence of the materialistic hype and shopping frenzy common in North America, but there was also much to be missed. There was little to invoke the spirit of Christmas and this was the first season we were without family or friends to celebrate with, not to mention the lack of our traditional goose and home made mince pies. Instead we went for Christmas Eve dinner to a recommended restaurant overlooking the harbour, where we had an enjoyable four-course dinner that catered to perceived fat gringo wallets.

On Christmas morning we were rudely awoken at seven o'clock by lewd and crude rap music blaring from our backyard neighbour's upstairs porch, where he and a buddy had already settled in for the day in full view and with an ample supply of beer.

To see in the new year, we accepted Mr. Gomez's invitation to a Rotary New Year Charity Bash at the Biltmore Hotel, at two hundred dollars a couple. That included a midnight buffet and a bottle of champagne at each table. Ours was a table of six, which included an American couple, also clients of Mr. Gomez. The bash started at eleven in the evening, a little on the late side for me, with a live and again very loud band, so it was hard to chat to one's table companions. We were determined to enjoy ourselves and with interest observed another aspect of Belizean society. This was obviously the upper social, business and political echelons of Belize and we were impressed by the friendly intermingling of the diverse ethnic cultures present. Most of the ladies were dressed in all their finery and the odd ex-colonial wore a dinner jacket. The buffet was enjoyable; rice and beans were included in the menu. Whatever the occasion, there was always rice and beans.

Although it was supposed to be the winter months, by our standards it was hot and humid. The locals were complaining of a cold front at 26 Celsius and most were bundled up in jerseys or jackets. An occasional sea breeze brought relief from the humidity, but we needed to keep our little fan on most of the time. Before many weeks had passed, it was badly rusted. We had not yet realised that plastic was the material of choice in the tropics. If the quality was good, it would last, otherwise it would crack in no time.

In his clothes cupboard Hugo discovered many of his leather belts and small leather items hidden under a layer of greenish blue mildew. We were amazed at how quickly these corrosive elements developed; none of which were mentioned in any of our readings about living in Belize.

Chapter 6

Travel and change of place impart vigour to the mind.

Seneca (54 BC-39 AD)

As we had no intention of living too long in Belize City or its environs, we soon started making forays into the southern and western parts of the country to get a feel for where we would like to settle. When we attended the Retirement Tour, we had liked the San Ignacio area in the western Cayo district, but upon visiting this area again, it did not say "hello" to us as it had before.

We were looking for a few acres of jungle that we could develop ourselves. Those we saw and liked were not an option as none had any water or electrical services available, even though posted signs read, "Coming Soon". We realised by now that coming soon in Belize could mean maybe never. We therefore turned our attention towards exploring the southeast that would take us down the Hummingbird Highway that starts in Belmopan and ends at the Southern Highway, the latter running just inland along the Caribbean coast.

A highway in Belize is a euphemism for a main artery road of which there are four: the Northern Highway, the Western Highway, the Hummingbird Highway and the Southern Highway are the only paved roads in the country, but they are narrow and poorly maintained. In total, only 575 kilometres of the 3,007 kilometre roadways are paved[7].

In the rainy season, when potholes are camouflaged by water, damaged tires or rims are not uncommon. Most of the unpaved roads are impossible to negotiate without a SUV. Even then, to go off the beaten track would mean sinking axel-deep in the sticky clay-like mud. Coupled with reckless drivers, driving in Belize can be a hazardous undertaking.

The Hummingbird Highway, dotted with small villages and settlements along the way, is where one sees some of the most beautiful parts of Belize, including real tropical rain forests. The hillsides are dense with majestic cohune palms which can grow up to 90 feet high and their fronds can reach about 35 feet, one of the largest of all palms. They occur naturally in the rain forests of Central America and have been an important source for sustaining life in the Maya culture. Their leaves are used for shelters of various kinds and the oil from the cohune nuts has many uses too, from cooking to lubrication. The seeds, fruits and heart of the palm are edible for both the Maya and their chickens, and wine is made from the sap of the palm heart.

Cohune palms are a little eco-system all of their own. Many ferns, epiphytes like bromeliads and even wild orchids can be found growing at the base of leaves. They also house forest birds and other little tropical critters. Dead cohune trunks are a favourite nesting place for parrots. They dig a nesting hole from the top, sometimes quite deep down. Unfortunately the local animal traffickers know this too, and one can sometimes see a series of holes made into the sides of the trunks, through which the nest robbers reach the chicks.

We were awed by the dense growth and varied flora we could see from the roadside. Not only trees, which have beautiful coloured blossoms in spring, but shrubs, ferns and different species of vines and epiphytes. In the small valleys we passed farms with acres of orange orchards. It was obvious which orchards were not well tended because wild vines shrouded the trees and were slowly but surely strangling the vitality out of them.

Here and there we saw clearings in the jungle, which are known as *milpas*, traditional subsistence methods of Maya agriculture. First they cut and then burn the original vegetation in order to plant their corn.

After having planted on the same *milpa* for a few years, the soil nutrients get used up and they then move on to a new patch and the process starts all over again. It does not take long for the jungle to reclaim the abandoned clearing.

Travelling south we passed through Stann Creek District, which is an important citrus and pineapple-producing region. By the changing vegetation we could tell when we were nearing the coastal flats, as the palms gave way to a variety of large leafed plants, including giant heliconias. We recognised them from having seen them in botanical gardens in Hawaii, but here they grow large and wild and one variety is aptly named Maya Gold.

Turning east on the Southern Highway, we reached the ramshackle coastal town of Dangriga. It is the home and cultural centre of the Garifuna, also known as Black Caribs, an ethnic group who proudly maintain their language and culture.

Briefly, in 1635 two Spanish slave ships from West Africa shipwrecked in the Caribbean and those who escaped and survived were able to land on the island of St. Vincent. There they intermarried with local Carib Indians and became known as the Garinagu or Garifuna people. They adopted the Indian language, but retained their African music and voodoo practices.

In 1797, when the island was under British control, the Garifuna along with other black slaves were relocated to the Bay Islands of Honduras. From there many settled along the coasts of southern Belize, Honduras and Nicaragua.

On November 19 1832, during a civil war in Honduras, a large group of Garifuna left that country by boat and landed in Belize at the mouth of the Stann Creek where they established the town of Dangriga. The 19th of November is one of the major annual public holidays in Belize, known as Garifuna Settlement Day. This celebration is commemorated throughout the country with Belizean food, Punta music and dancing, the latter that includes "jump up" (just as the name indicates), a reenactment of the landing at Dangriga and just another reason to party.

To us Dangriga seemed like a micro Belize City, not as crowded but more dusty. From the creek mouth one can hire a water taxi to go to one of the numerous cays off the coast. The taxis are mostly narrow open fibreglass boats with hard wooden benches and an outboard motor. A hat and a cushion are advisable.

We fell in love with the Hummingbird area and made frequent trips that way and further south.

The Hummingbird area

Hopkins is the next village to the south of Dangriga, a quaint and scruffy little Garifuna settlement by the sea. A few miles south of Hopkins is a small resort area we wanted to check out.

On one of our flights to Belize I sat next to an elderly Canadian man who said he was on his way to visit his daughter and son-in-law who owned Hamanazi Resort. They had scuba diving facilities, which were of interest to us. We found Hamanazi to be an attractive little resort, nicely maintained and in keeping with its location. Our lunch was not worth raving about, but we felt they did a good job considering their remote location. It's an ideal place to have a holiday far from the madding crowd.

Even further south was the town of Placencia; another laid back little village whose reason for being we could not quite see, except that it was even further away from everything. The stretch of road to the village from the Southern Highway was about twenty miles of appalling dust and potholes, which we could image to be a nightmare in the rainy season. There were hardly any shops and other necessary facilities close enough in that region. To drive all the way to Belize City or even Belmopan for shopping would be a whole day's undertaking, not to mention what we would do if we had a medical emergency.

We had to ask ourselves how would we be affected if a hurricane hit the coastal area. I could not see us huddling in those inadequate looking hurricane shelters with our pets. On our first trip there, we decided to rule out any place beyond the mid-point of the Hummingbird. We had come to Belize for a restful, not a stressful retirement.

***

On one of our earlier reconnoitres along the Hummingbird, we had noticed a "for sale" sign by a ridge along a lovely part of the highway that really appealed to us. The jungle had been partially cleared, revealing a mix of orange trees and cohune palms on the lower mountain slopes. We decided to investigate. Our enquiries along the road led us to a Mr. Polanco, who lived just a short distance away. He was the owner of one of eight privately owned bus lines whose "chicken buses" (as they are known to gringos) were only licensed to operate in the rural routes of the southern region.

His yard looked like a real junkyard with numerous old bus bodies lying in between his still-operational one. I assumed he needed to have spare body parts to keep the current bus in operation. That would be necessary, considering his route did not have many paved roads. After calling out a few times, a squat looking man of mixed Belizean ethnicity appeared from behind the bus carcasses. He reminded me more of a yesteryear Caribbean pirate, with a bandana covering his broad head. We were told the bandana was to hide the permanent evidence of a machete chop to the head.

After exchanging friendly greetings and explaining our quest, Mr. Polanco told us to follow him. His much younger common-law wife (as we presumed her to be since most unions amongst the locals are "common") and his adult son also climbed into his lorry and we followed them up the Hummingbird and stopped at the top of the ridge. The wife pulled old nylon stockings over her arms for protection against the biting midges. Polanco seemed to be monarch of all we surveyed as he gestured to property on either side of the highway, from which we could make a choice. We explained that we were only interested in two or three acres, preferably where we had seen the for sale sign.

Off we went again and turned onto a grass track that went up and onto the cleared ridge that had drawn our fancy in the first place. It was breathtakingly beautiful, covered with grass and dotted with various palm trees and a view over the valley and the opposite mountain slopes. Citrus trees grew on the southern slope that led down to a small gulley. The northern slope was covered in natural growth with a small stream at the bottom. The highway was hidden from view by more dense growth.

We walked higher up the ridge, where with his machete, Polanco knocked down and opened a few green coconuts. (A machete is to a Belizean what a pocket knife is to a gringo). He offered us the coconut water. Hugo drank heartily but I did not like the bland taste.

Hugo drinking fresh coconut water

At that green stage, the coconut flesh is still soft and almost jelly-like and not good for eating, unless one needs a laxative. Green coconut water is a favourite local non-alcoholic beverage and only ripe coconut water is referred to as milk.

Polanco pointed to the mountain peaks higher up the ridge, explaining that was his water source and we would only need to connect our pipes to his main pipe. For electricity, we would most likely need one more pole to connect to the lines along the highway. In my mind I could already see us on this lovely patch of tropical paradise.

Back on the highway and closer to the bus yard, we stopped at another spot that was on the market. It was not as appealing as the one higher up. A cluster of huts that housed all Polanco's adopted children, was too close for our liking. Someone later said that the purpose of the children was to provide labour on the Polanco farm.

We were eager to pursue this property negotiation. The price was right and he assured us that he held the title to the property. The only obstacle that needed to be cleared was for him to get a surveyor to measure off three acres, and then attain approval from the relevant government department to sub-divide. We just needed to wait.

As we were spending so much time and energy driving away from Belize City, we decided it would be more practical to rent a house in Belmopan to be closer to the areas we were scouting out. When we had made a quick bus ride through Belmopan while on our Retirement Tour a year earlier, the town appeared quite uninteresting, dusty and drab. Moreover, we had read in Lonely Planet guidebook on Belize, that the best thing to do in Belmopan is to take the first bus out.

However, by now we had come to accept the dishevelled look of most things created by man in Belize, and besides, we were only considering renting in Belmopan. Compared to Belize City, Belmopan was a sleepy hollow and way less squalid and ramshackle. It would be a good move.

Shortly after our decision to move was made, on one of our frequent trips to view the coveted patch of jungle on the Hummingbird, we pulled into the shade just outside Belmopan to eat a take out lunch. At the Belmopan open market we had found a little eatery where one could buy take out Belizean food. It was a place that would definitely be condemned by health authorities in Canada, but in Belize one needs to be less fussy and hope for the best. We were told their food was good, so we bought barbecued chicken with rice and beans.

While eating, a small lorry stopped in front of us and the driver walked over to ask if we were from South Africa as he saw the international

ZA identity sticker on the back of our SUV. In a familiar South African accent he introduced himself as Paul, an expat originally from Durban, and invited us to visit him at his nine-hole golf course next to Roaring Creek, a few miles west of Belmopan. We promised we would.

In the meantime we waited and waited to hear from Polanco. Ernesto told us that in Belize one needs to use patience, and if one does not have any, one will learn it. How true that was. To get a progress report on the property negotiations, we had to drive the two hours from Belize City to personally speak to Polanco as he proved unreliable about phoning us. He said he was also waiting, as he had submitted all the necessary applications to the Lands Department.

He assured us that we could go ahead with our building plans anyway. That we would definitely not do until we had everything signed. We had read enough dire warnings about gringos innocently buying property from Belizeans, only to find after the deal was done that the seller did not have clear title.

On one of our trips to the Hummingbird, while walking around and visually planning on the coveted property, our ankles came under attack from the most noxious of all the many biting midges in the jungle; one of the many species of phlebotomine flies that are almost invisible, live in the grass and attack any penetrable warm-blooded surface that disturbs their habitat. Like other vectors, it is the females who bite and suck blood and the usual brands of insect repellents do not provide protection. Thin gringo skins, not toughened by the outdoor elements, are especially susceptible.

The bites leave little pinprick sized blood spots on the skin that itch for weeks. On this occasion Hugo was particularly badly bitten and his lower legs looked as if he had a severe case of the measles. We discovered if we immediately squeezed out the little blood spot where the bite occurred, the itching stopped fairly soon. Obviously that was where the toxins in the saliva were deposited.

***

Shortly before leaving Canada, we were given the names of Anne and Noel, a Canadian couple who used to live on Vancouver Island and were now citrus farming along the Hummingbird Highway. One day we realised their address was not too far from our coveted patch and as we did not have their phone number, we stopped by for an unannounced visit. They were in their mid-seventies and had moved to Belize about ten years prior to us. They had built a plain, but adequate concrete house with a screened in porch on two sides.

I was surprised that they did not have security bars on their house, but two mixed breed dogs (known as "pot lickers" in Belize) were supposed to be their protection, even though they were always confined to a corner of the porch. Their house, situated on a slope a short distance from the highway, afforded them a lovely view of the valley. It seemed like an idyllic existence.

They told us about Polanco's brother who had initially befriended them, and had offered to undertake on their behalf certain financial transactions they needed to make from Canada before moving to Belize. In the process he also took care of some of his own transactions with their money. We had read many warnings not to give Belizeans authority over one's money. Although promises were forthcoming every year from the relevant departments, Anne and Noel had not yet received telephone, cable TV or high-speed Internet: all the services available throughout the country, according to Bill and Claire Gray's Retirement Guide.

We soon became friends and benefitted from their advice and experiences. It was Noel's opinion that acquiring the services of a reliable builder in this remote area would be difficult and more costly. After much waiting, and driving back and forth, Polanco informed us that he was having difficulties getting permission to subdivide. Fortunately we realised by now that he would not need permission to subdivide if he owned clear title and the land therefore must have been leased land. It seemed that defrauding gringos was in the family genes.

He offered us a different patch of jungle, which he did own outright, but it was not to our liking and by now we were having second thoughts about the whole Hummingbird enterprise. As lovely as the area was, I was definitely not going to settle where there were no communication facilities. There was no guarantee that any would soon be forthcoming either. We decided to search elsewhere.

***

A week after meeting Paul, we took him up on his invitation to visit him at his golf course. It was situated at the end of a bumpy dirt road a few miles off the Western Highway, not far from Belmopan. We were pleasantly surprised to see such a lovely place in the middle of the bush. Apart from the expansive lawn and patches of beautiful tropical shrubbery, a pond attracted various birds and a little stream was home to a small crocodile.

Upon arriving at the house, whose front area served as the clubhouse, we were greeted warmly by a black man who introduced himself as John from Zimbabwe. After fleeing Robert Mugabe's Zimbabwe without a passport, and the ensuing difficulties, he and his family had eventually been given safe haven in Belize and he was now in Paul's employ.

Our dogs had been on their leashes for most of the time since we left Canada. Now, with newfound freedom on the expanse of Paul's lawn, they ran around and around until they were exhausted. When we left the golf course, I noticed they were teeming with fleas picked up from the lawn via Paul's "pot lickers". Fortunately Hugo had given the dogs their regular dose of flea treatment the day before as no place in Belize is safe from a host of parasites, so the fleas died off quickly.

In our search for a rental house in Belmopan, I had come across a phone number on the Internet. The person who answered had a Canadian accent and introduced himself as Dewey. He said he was only a contact point for a Belizean realtor named Yolanda who didn't have a phone. We arranged with Dewey a time and place to meet with Yolanda in Belmopan. He also told us where he and his wife Lilly lived and invited us to visit. Therefore, off to Belmopan we went for the umpteenth time. We met Yolanda at the appointed time and she took us to a little concrete house that looked identical to all the others in the vicinity, aptly referred to as "cookie cutter" houses.

They were all within a stones throw of each other; all situated on a patch of bare ground without a blade of grass or a tree in sight and no fences around. The rent was BZ$1000 a month, whereas our rental at $800 in Belize City was a mansion by comparison. The next house we looked at was similar, but in a better part of the city. It had a lawn but no fence. When we walked around the house, we had a whiff of that by-now-familiar sour sewer smell and the lawn squelched under our feet where the septic tank was overflowing. We declined all offers.

Our next stop that day was back to the golf course. We made sure to keep the dogs in the car this time. Paul and his (then) Belizean girlfriend Jackie told us that those cookie cutter houses Yolanda had shown us were being rented to Belizeans for $450. We had again experienced another attempted gouging. Jackie said she knew where we could rent a half decent house and arranged for us to meet the owner.

Chapter 7

There is no pity for a man who moans about living in one town and does not move to another.

The Talmud

At the appointed time we followed Jackie to a house near to the cookie cutter house with the overflowing septic tank. Shirley, the owner of this house, was waiting for us. She informed us that she had been a city councillor and was on staff at the University of Belize. As we understood, for financial reasons, she rented out the house after she and her husband parted ways. She and her daughter were living with her mother and were within walking distance of the university. The present occupant, also on staff at the university and whom she constantly referred to as "the doctor", was due to move out soon.

This house would not qualify for *Better Homes and Gardens*, but it seemed perfectly adequate as a temporary abode. It had been a cookie cutter house at first, but Shirley had improved it by adding on a spacious bedroom to one side, creating an L-shaped house, which allowed for a covered carport. The short driveway was made of concrete, there was a rainwater tank near the back door and the whole yard was planted with grass and completely fenced.

Rental house on Barbados Street

All of that satisfied our housing needs and we happily agreed to $700 a month. Jackie informed us later that Shirley had wanted to charge us more, but she persuaded her against it. As her tenant was moving out, she had little choice. Gringos looking to rent were few and far between at that time.

I liked Jackie. She was a very classy Creole girl who visited me frequently. She made a point of warning me about young Hispanic girls who used their wiles to captivate vulnerable gringo men in mid-life, in the hopes of getting their hands on the man's presumed wealth. Once the girls had accomplished their mission, they moved on to new pastures. Fortunately my husband had passed his mid-life, but we came to know a few nice gringo men whose marriages had succumbed to these self-inflicted wounds. Jackie moved on to better opportunities in America.

In mid-February 2004, we moved to Belmopan.

Living in Belize City for two and a half months was an interesting experience, but we were happy to leave behind that frenetic city. The same gang who had unpacked our container, moved our goods to

Belmopan in two delivery vans. We followed behind the last van, where the gang had arranged our living room furniture and made themselves comfortable in our lounge chairs, causing me to have fits as I noticed their pants leave sweat stains on my cushions.

Leaving Belize City

Our rental house on Barbados Street was located next to an empty corner lot in middle class Belmopan suburbia, just one street away from the better area that housed the well-established British High Commission and a few other embassies that had recently relocated from Belize City. The Brazilian embassy residence was a beautiful newly completed rental house, reputedly built with drug money.

Some of our neighbours were educated civil servants, who took pride in their little concrete homes cheek-by-jowl with each other. Their various domestic activities were quite audible from early dawn. Barbados Street, like most others in the city, was unpaved and dust settled everywhere.

In Belmopan we encountered the same problem as we had in Belize City of being required to put down a small fortune in order to get telephone and Internet service. A friendly bank manager suggested an alternative Internet service provided by a gringo security company, which would only require an antenna on our roof.

Our life in Belmopan became closer to the norm and we were able to unpack more of our household goods. The added-on bedroom had a small anteroom, which served as our office. The second bedroom was just large enough to fit a queen-sized bed and two night tables on each side and the addition of a portable zip-up clothes hanger. The third room was used as storage for all our unpacked boxes and furniture.

For the first time our dogs were able to go out freely into a yard without the danger of picking up undesirable parasites. We were surrounded by bird life of all kinds, including parrots and parakeets flying noisily overhead each morning. A pair of kingbirds made a nest under the eaves and we, and the cats that were housebound, had a front row view of their activities, from the laying of eggs to the chicks fledging the nest.

The lawn had many round holes in which tarantulas lived. Their burrows were lined with silk and they came out at night to hunt, so fortunately our paths did not have to cross too frequently. Fortunately too, no cockroaches came into evidence. We frequently found scorpions in the bathroom, so we had to check our towels before we used them. I knew they liked water, but it remained a mystery how they found their way to the bathroom, since all the windows had bug screens.

We soon realised we had unknown tenants above our bedroom ceiling, as we could hear the pitter-patter of little feet at night, and at times lots of squealing. There was an obvious entry hole high in the wall from the carport side, but we were not sure whether they were rats, bats or possums; the latter we occasionally saw lurking in the carport some nights.

Our neighbours told us they were rat-bats. Whatever they were, they were responsible for some cruddy stuff of undetermined origin that drifted down the walls from the gap where the walls and ceiling met and settled on the furniture against the walls. To solve the problem, we decided to stick masking tape all along the wall where moulding should have been. The masking tape would not stick and we realised that the house was not painted with regular paint, but had been whitewashed with a low cost type of wash made from slaked lime and chalk. That became more evident when wiping a mark off the wall: the spot become darker instead of disappearing.

***

After spending more time driving around the country and looking at property, we realised that any property off the beaten track would be a major undertaking that we were not prepared for; we would have had to provide our own water and other utilities at great expense. Therefore, we decided to settle for the Belmopan area that by now was familiar and we liked that the city was relatively quiet and centrally situated in the country, being about fifty miles from most important points.

On an Internet search we found Jimmy, a realtor who fortuitously lived in Belmopan. We arranged a meeting with him and learned he and his family originally came from Taiwan. We found out too that we shared a common Christian faith and therefore felt confident in hiring him as our agent. On the outskirts of the city, but still within city limits, Jimmy showed us five one-acre lots shaped like giant dog runs which belonged to other Taiwanese. We decided on the one second from the corner.

In Belize one's realtor can do transfer of ownership and accompanying legal formalities at the Lands Department. Moreover, it was especially advisable not to go through a Belize lawyer unless one has been recommended by a reliable source. The lands office and many local *laayas* (lawyers), as they are so aptly called in Kriol, are notoriously corrupt. We had first hand knowledge of a less astute Haitian-American couple from Boston in the United States who were cheated twice. The first time someone who was not the legal owner sold them a piece of property. The court ruled that they could be refunded when the legitimate owner eventually sold the property. There was a slim chance of that happening.

Next they bought a piece of property through a local *laaya,* believing they were protected this time. The seller assured them that the quoted price was in Belize dollars. When they went to finalise the transaction, the lawyer insisted the price was in US dollars, which meant double the Belize price; the same trick used by the checkpoint official who had fined Hugo for me not wearing a seatbelt. We were pleased to read that he had been arrested after accepting a cheque from another victim, which he then cashed at a petrol station.

There are constant reports in the media of corruption in the Lands Department and in the Cabinet. There are allegations of land deals done without going through proper channels and of cabinet ministers giving themselves or their relatives valuable land even on the cays when they realise that they could lose an upcoming election. Allegedly some have even received leases and title for portions of seabed without approval from environmental departments.

Ordinary Belizean citizens can apply for land. After they have proved development, such as growing crops or cultivating fruit trees for a stipulated number of years, can they apply for title: until then it is regarded as leased land. Allocations are frequently made on the basis of political cronyism or family connections.

In Belize there is almost no bringing to book any politician as most of them have had their hand in the cookie jar at some time or other. Every now and then citizen groups would stage protests to bring attention to the inequality of land distribution.

Fortunately, our property purchase was concluded without any problems or delays. We became formally friendly with Jimmy and his family over time and Hugo tutored his daughters in biology and chemistry. After the girls graduated, the family moved back to Taiwan.

***

As non-Belizeans we had to pay a 10 percent "purchase tax", and after that a low annual two-figure property tax. With this important step concluded, we could settle down and plan our future in Belmopan.

The first step was to clear our property. It was covered with thick jungle vegetation of all kinds, including some vicious-looking thorny vines called *cocolmeca*, whose stems grow up trees and twist into hardwood knots around branches. The main bulbous root is regarded as medicinal from which the locals make a tea, believed to cure tumours, leprosy, psoriasis and just about any other ailment. The only way to destroy the vine is to cut off the stems and wait for the knots to dry out sufficiently until they can be broken apart. Other vines were long and soft with tiny, almost invisible hair-like thorns that burned ones skin and were hard to extract. One long established expat acquaintance had warned us that everything in Belize either bites, burns, stings or steals.

We had to use machetes and a compass to clear our way onto the property and find the boundary markers. A local man who saw Hugo hacking away at the bush offered to do the clearing for a fixed price which included raking into burn piles. Hugo gratefully accepted and a gang with machetes duly arrived the following day. They cleared our acre and the acre next door and insisted on being paid for both. Hugo held his ground that he was not financially responsible for their dubious mistake. They did not return the next day and Hugo ended up doing all the raking and burning himself.

We were also getting to know our way around the town. Apart from a few not-so-tidy-looking small Chinese-owned stores, Belmopan had one small supermarket that was Belizean owned and not that super. They catered to the gringo market and stocked a limited variety of items from America that were familiar to us, but at inflated prices. Some hardware items were double that of identical ones at other stores. Even pointing out this discrepancy didn't make any difference to them. It was therefore not a surprise to learn a political family owned the store.

Then I experienced what other gringo women had warned me about: not to purchase any dry food items as the store was contaminated with weevils. I was pleased to find that the store did stock white baking chocolate, only to find that it was full of tiny little wriggling pink worms. I never imagined that one would find worms in chocolate. After that I always inspected the inside of the packaging before purchasing baking chocolate and many other items. The telltale evidence was in fine web-like threads in the box. Once back home, the items went straight into the refrigerator.

In spite of the limited variety, our basic needs were met and we only lacked non-essentials such as fresh cream and chocolates. Being a chocoholic in Belize was not an easy position to be in. Whenever we had an occasion to go to Belize City, we stocked up on decent butter and other goods that were not available in Belmopan.

***

Fresh fruit and vegetables were available at the regular open *mercado* (market) under roof, mostly run by Hispanics. It was difficult getting parking there as the spaces in the front were mostly reserved for the officials from the government buildings close by. I got into the habit of questioning whether the quoted price was for gringos or locals, as some merchants pushed up their prices when they saw a gringo approaching. Since we were retirees lured by the promise of inexpensive living, I took quick umbrage at being gouged.

The *mercado* also had "eateries" where locals could buy prepared food. In a speech during Prince Harry's visit to Belize early in 2012, the mayor of the city promised to build a new market, as the present one was too old and too infested with cockroaches to even attempt a clean up. Behind the market was a small freestanding wooden hut that housed a "restaurant" with typical Belizean food and owned by a friendly Creole lady. There were about five small plastic tables and chairs for indoor dining. The kitchen was a very narrow corridor-like section at one end and had no refrigerator. A plastic-covered wooden shelf with open shelves underneath stood against the wall and served as a work counter. In the narrowness of the kitchen the three large ladies who prepared food could not pass one another at the same time.

The place had a reputation for good local food and the menu was varied. The chicken was barbecued outside by the husband on a typical Belize barbecue that was made from an old butane tank cut in half lengthwise; the bottom half held the charcoal and the top half served as the lid, all welded together on a rebar frame. Apart from rusting, they worked efficiently and we came to own one too.

A choice of barbequed or "fry chicken" and rice and beans or "stew beans" were always on the menu, along with boil-up (a boiling-together of all the root vegetables with a pigtail added, buckets of the latter coming from Canada), cow-foot soup, which is a self-explanatory traditional favourite and other more familiar-sounding choices.

When waiting for an order I had to exercise mind-over-matter and I never ventured beyond the barbequed chicken with rice and beans, but Hugo was more adventurous and tried everything except the cow-foot soup. We frequently took take out from there when we had to provide any of our workers with a meal. One day Hugo and our weekly gardener, who regularly had lunch together on our stoop, both felt the chicken tasted off: after that we did our own cooking.

***

On Tuesdays and Fridays a colourful "entrepreneurs' market" sprang to life on an open area not far from the regular *mercado.*

Entrepreneurs' market

There one could buy almost anything that was to be had in Belize. Mennonites came all the way from Orange Walk north of Belize City, with their trucks piled high with their simply made hardwood furniture that was purchased mostly by the locals. Taiwanese sold junky trinkets from China, as well as Phalaenopsis orchids, seasonal yams, bok choy and other leafy greens not commonly grown by the local gardeners.

Local merchants sold vegetables and fruit in season, which they purchased and loaded up the evening before from Belizean farmers who came to town to sell their produce.

Hugo checking the vegetables

Their selection was limited to vegetables that could survive the hot and humid tropical climate, which ruled out lettuce and other leafy salad-making foods.

Mexico had fruit and other fresh produce available year round. Apart from a few items imported by special permit, I could never understand why the Belize government did not allow Mexican fresh produce to be sold locally when Belize was out of season.

Frequently a gringo-gone-native-with-blond-matted-dreadlocks sold second hand books and other bric-a-brac at the market. One could also buy pirated movies, albeit of a poorer quality, as Belize does not honour copyright laws. A colourful local jungle doctor occasionally set up her own stand where she sold bottles and packets of her homemade potions that "cured all ailments" from various plants she collected in the bush, some specifically labelled for women and others for men.

Janice, the jungle doctor

Belizeans are firm believers in jungle medicine, thinking that if it's "natural" it's safe. One gringo farmer we knew had a worker bitten by a *fer-de-lance*, or yellow-jaw as the locals call those very venomous snakes. He refused to be taken to the hospital and insisted on jungle medicine: he died. Less understandable are the many new age and other holistic-leaning gringos attracted to the rainforest, who also believe in jungle medicine and are quite confident in swallowing untested products rather than medicine whose dangers and benefits have been well researched.

Surprisingly there was a store selling Birkenstock shoes located in a converted shipping container fitted with air conditioning. The owner was a German architect who had lived in Tanganyika for twenty years (now known as Tanzania) before moving to Belize. He was a proponent of reflexology and believed that one's feet were the source of all ailments, so it was not surprising that he promoted healthy shoes. I eventually came to own a pair and they were the only shoes that survived the rigors of Belize.

Perhaps the majority of merchants were those who sold second hand clothing.

Second hand clothing market

Like all other vendors at the entrepreneurs' market, they would arrive early in the morning with their vans filled with clothing; set up their collapsible awnings and unpack their merchandise. Some were organised with railings on either side where the clothes hung in neat categories. In the centre was a trestle table piled high with folded clothing, though very soon the piles would be a mess. Others were quite disorganised and items would be piled on a table or on plastic sheeting on the ground.

In some of the poorer outlying neighbourhoods of the city, one could also find clothing for sale. These were laid out on plastic sheets on the ground and were known as "Bend Over Markets".

Not all vendors carried the same line of goods. Some sold stuffed toys and children's clothing; some only male or female items; others sold mixed genders. There were not too many used shoes sold and I never saw second hand underwear. These latter items were newly manufactured in neighbouring countries like Guatemala or Honduras. Belize City was the only place where one could find some new clothing stores, but they catered to the local fashion styles and tastes, which definitely did not follow the European trend, especially for women.

***

This trade in second hand clothing fascinated me, but I could never find out from any vendors exactly how the system worked so I went to that source of all information, the Internet. The second hand market seems to be a multi-million dollar business originating in the First World and ending up in the Third World. Ninety percent of the unwanted clothes we give away for various reasons to charitable organisations and businesses are sold to recycling textile firms, from where much is recycled into industrial cloth items. Twenty-five percent of the remaining clothes are sorted into various categories and classifications, A1 being the best quality. They are then packed into sealed bales generally weighing between 100 to 120lbs and sold again to other agents who ship the bales to different poorer countries around the world.

Some bales contain only blue jeans, or other specific articles while others contain mixed items, all at their own particular pricing by weight. The bales shipped to Central America normally come from North America. A vendor told me that they choose what type of bales they want to buy and I am sure they share their bales with others as I never saw any vendor who had 100lbs of one item. What was originally an expensive pair of jeans ends up costing a local about ten dollars, a bargain for them.

I read that clothing from the east coast of America was preferred to that of the west coast as the former's clothing was in better condition. That sounded feasible, since the east coast is more "establishment" and they don't wear their clothes out to the same extent as west coasters generally do. The west coast is after all the birthplace of grunge.

There is a downside to this industry in poorer countries, especially in Africa. Even though it creates affordable clothing for many, it also destroys local manufacturing. Small local businesses cannot compete with a cheap second hand market, thus creating unemployment. Often our good intentions are not so good after all. I did not notice this downside as being a problem in Belize, as there seemed to be no small local industries to speak of. The few that did exist were part of the "fair trade" market and their goods were sold abroad.

***

We were surprised at how cosmopolitan Belmopan turned out to be. Lilly and Dewey, some of the first people we met (our contact for the first realtor, Yolanda) were well-travelled Canadian survivors of the sixties generation; VW bus and all, who eventually found themselves in Belmopan. Lilly runs a busy small Bed and Breakfast as Belmopan has a serious shortage of visitor accommodation.

Lilly in turn introduced us to Peggy and Dick, who were expats from the United States. Dick had been a park ranger in Hawaii and retired to Belize because he wanted a patch of rainforest to call his own. They bought a large piece of property a few miles from the village of Teakettle, which is not too far from Belmopan and they had a lovely view of the Belize River. Being a disciple of Darwin, Dick kept half of the farm natural and on other parts he cultivated "organic" pitayas, more commonly known as dragon fruit on the North American market.

Peggy in turn introduced us to Audrey and David, an expat couple from Britain. David had taken early retirement from the Royal Engineers. Audrey grew up in Belize, but moved to Britain at an early age to study nursing and then joined the British army, where she met David. With his skills they built a nice house on a hill near the village of Blackman Eddy, and had a lovely view overlooking the large Mennonite community of Spanish Lookout.

Dick, David and Hugo enjoyed outdoor activities together, most notably a three-day kayak trip down the remote Raspacula and Macal Rivers for which they had to obtain a special permit. They were enthralled at the sight of so many pairs of scarlet macaws, which contradicted the naysayers who predicted their demise because of the building of the controversial Chalillo Dam.

Getting to that area and back was an adventure in itself, as the road down to the river comprised mostly of rocks, some quite large. There our Nissan Xterra showed her true mettle as she successfully bounced from rock to rock. Later David developed a respiratory problem that no medical facility in Belize could diagnose, so he had no choice but to return to Britain where he was diagnosed with terminal mesothelioma. Audrey eventually returned to Belize alone. We felt the loss.

On our way home to our rental house, we used to pass a house under construction in the vicinity of the British High Commission. As we were soon to embark on our own building project, we were curious to see how things were done and stopped to look in. That is how we met Steve, a transplanted Canadian and his Venezuelan wife Fabiola, owners of a successful citrus farm and a few other enterprises.

Chapter 8

All the art of living lies in a fine mingling of letting go and hanging on.

Henry Ellis (1859-1939)

During the spring of 2004 we were busy planning our own building project and working on plotting and preparing our patch of jungle. From a home-planning book I modified a beach cottage plan to suit our tropical location. We were put in touch with a municipal building inspector who was moonlighting as a draftsman and for a fee drew up our blueprints. The local grapevine was also at work and before long Hugo was accosted in various places by an assortment of unknown locals offering their services as builders or contractors. Many seemed suspect right off the bat.

Only a few understood the concept of cost per square foot. They put down what they thought the cost of material would be; added more for "contingencies"; doubled that to cover labour costs and voila, there was the quote. Hugo asked for a quote from one man who sounded fairly knowledgeable. Knowing of course what the average cost per square foot was, Hugo could tell that the quote he presented was way out of line. When Hugo turned him down he was quite angry and demanded to be paid eight hundred dollars for the quote. He had no choice but to be satisfied with fifty.

By this time we had been in contact with Chuck and Judy, the expat couple from Oregon we had met on our Retirement Tour. As Chuck was a building contractor in Corozal, they walked us through all the information needed on "how to build in Belize", for which we were grateful as it saved us from a lot of potential trouble during the project. Eventually we received a reliable recommendation for two brothers of Maya Mexican fusion who were builders and lived in Succotz Village, not far from the Guatemala border. They just happened to be working on a house not too far from our rental house, under the supervision of the German architect who sold the Birkenstocks. We could therefore see their workmanship, and before long Hugo struck a deal to everyone's satisfaction and signed a contract with them. They would start building for us when they completed their present job. That gave us time to look around in the limited hardware stores to see what materials and fittings were available and to our liking.

While we were waiting to start building our house, we had other things to attend to as well. Despite the fact that neither Hugo nor I had very green thumbs, we looked forward to the adventure of creating a tropical garden. Many plants were familiar to us from my mother's lovely garden and other parts of the sub-tropics of South Africa, as well as what we saw in the botanical gardens of Hawaii. We were particularly keen to grow our own fruit trees, to which end I had ordered seeds before we left Canada. Hugo knocked together a stepped nursery stand and we started some bedding plants.

During the summer months when the mangos came into season, we planted pips of some mangos that we found particularly delicious, namely a "Cambodiana". Those are an early variety and the only ones whose name the green grocer vendor knew. We were told that we needed grafted trees in order to get the true variety. We would see what our pips produced. Most of the mango and also avocado varieties one bought at the market were unknown to the vendors. It all seemed rather hit or miss.

As we had no adequate garden equipment as yet, we were happy to use the services of an itinerant gardener while in our rental house. There were enough Hispanic men on bicycles, with lawnmowers and other rudimentary equipment perched on the handlebars that sought out regular clientele. One had to be careful that they did not add any of one's own equipment to their supply. We had to replace our rakes every now and then.

Our portable cement mixer, stepladders and mountain bikes were all chained to the carport posts. In addition, the gate to the yard was always locked. With the first rains of the season we discovered that the lovely looking green lawn, indigenous to Belize, did not form a thatch; instead the leaves just lay on the ground, hiding red mud that squelched through onto one's shoes or between one's toes. Mud splatters on one's clothing created permanent stains from the iron oxide in the soil.

We decided to get two big black dogs, as people had repeatedly said that they were the best security. Someone told us about German Shepherd puppies for sale by a missionary in the adjacent rough village of Roaring Creek. That village could benefit from a missionary or two, as it was renowned for having the highest per capita crime rate in all Belize. We preferred the idea of well cared for puppies from a gringo breeder to the unknown parentage of litters born in indifferent circumstances.

We felt reassured when we arrived at the missionary's property to see healthy looking dog parents with sweet little puppies safely contained in a large fenced yard. The puppies were being sold at six weeks instead of a healthier eight weeks so we picked two, knowing it was an advantage for their development to have companionship of their own kind.

Once we arrived home we realised they were teeming with ticks of all varieties and stages. As they were too young to be treated, we sat for hours picking off the parasites. However, by that stage the yard was contaminated and we had to think of tick extermination. The acaricide Hugo used necessitated weekly spraying of the entire yard for a month. After spraying, the dogs had to be kept off the grass for a day, then the grass had to be well watered; after all that it was safe for the dogs to go outside again. It was quite a palaver.

As April drew nearer, the heat and humidity increased. The little fan we had bought in Belize City was completely rusted. Hugo didn't seem to identify with my discomfort and after much nagging on my part, he eventually agreed to a larger and more robust one that was also noisy.

Even though the fan was positioned at night to blow directly on the bed, I still found it necessary to wet a bath towel I then laid over my body. By the time the towel was almost dry, the air was generally cool enough to enable me to fall sleep. Hugo never seemed to complain about the heat.

About this time the foundation for a new American Embassy started not far from the British High Commission residences, and all the dump truck traffic was diverted down Barbados Street and past our house. The dust from the street settled in a thick layer on the furniture; therefore, I was compelled to keep the front windows closed.

***

News and gossip travelled fast in Belmopan and it did not take long for locals to know there were new gringo settlers. It is an unwritten rule in Belize that no one just walks into one's yard, unless they have ulterior motives. They blow their hooters outside the gate and if they don't have a vehicle, shouting does the job.

We had various people come to the fence to solicit money for some reason or another. A common scam was a woman with an official looking albeit well creased piece of paper, ostensibly giving her permission to collect money because her house had burned down in Belize City. After living in that city for a few months we had learned how to recognise some scams. With this particular one, I wondered how this person could afford to come all the way to Belmopan to collect money. She did not go from door to door either, but zeroed in directly on the new gringos' house.

One day as I was hanging out laundry, a man walked across the open corner lot directly to our fence. By now we had been told that Belize men would not directly ask a woman for a loan. When he asked to speak to my husband I knew what he was after.

He told us he drove a fish truck that had run out of petrol about twenty miles east of town and he wanted to borrow money to fill up before the fish went rotten. When we asked him how he had gotten to town without his vehicle, he told us his cousin had fetched him. We suggested he borrow money from his cousin instead of from us who did not know him. We were always amazed by the imaginative stories people thought up to scam newcomers out their money.

The summer months are supposed to be the wet season. Fortunately for our building plans, this year was not the norm and little rain fell. The season's rain seemed to have been stored up for one particular day in early July when the skies burst open and rain poured all day; so much so that windshield wipers could not work fast enough to allow for visibility when driving and we were forced to cancel an arranged visit to Anne and Noel down the Hummingbird.

At the end of that day there wasn't any water coming out of the taps. In Belize it takes a while before any information or explanations are forthcoming. After a number of days we learned the water mains had washed away in the rain and it would be ten days before we could expect water service to be restored.

Fortunately we had a rainwater tank outside the kitchen door. Our pioneering days in South Africa kicked in and we set up bucket stations in the bathroom and kitchen. It took many daily trips to the water tank but we managed without too much hardship. It is possible to have a thorough wash with a bucket and a jug. Even though there were fairly frequent interruptions in the water service, it surprised us how many homes in Belize did not have rainwater tanks.

On occasion we invited our landlady for a simple dinner and learned a lot more about Belizean customs. Most Belizeans generally drink locally made beer or rum. Wine and spirits are imported and therefore expensive as most of the country's revenues come from import duties. Moreover, the wine selection was rather poor. We therefore only kept a few bottles of better wine stashed away for special occasions, which did not include entertaining the landlady.

The first time she came for dinner she brought a few additional unexpected guests. We offered beer and soft drinks all round but she asked if we did not have wine. Not being fast on my feet to tell a lie I said, "Yes." Therefore, we had to bring out a bottle. After dinner I was asked to pack up some of my very-much-enjoyed food for her to take home. We were rather taken aback by that request, but learned soon enough that that was how things were done in Belize. Understandably it was a situation I avoided where possible from then on. On another occasion our landlady asked us to fetch her as she had been to see the doctor, who had ordered her not to drive for six weeks because her "womb had dropped". We never found out how such a medical condition could occur. She shared a Toyota lorry with her son and she had explained to us what a serious financial outlay it was for a Belizean to buy a new vehicle. Perhaps it was the son's turn to use the lorry.

With time we noticed an obvious increase in the number of expensive SUVs and other fancy cars on the roads, especially those driven by younger men. Knowing what our landlady had said and what the average income was for Belizeans, most of those could only have been acquired through political or drug connections.

***

When we moved to Belmopan one of the first things Hugo did was to go to the local police station to submit all the necessary applications to register and license his guns that were still in Queen's Custody. He was taken to the assistant commissioner of police who gave him the once over and said, "We will let you know." Six months later there was still no word from the said official, so Hugo turned to our landlady for help.

As she had political connections, she arranged for a meeting with the police commissioner, whom she knew personally. So there it was, not "what" you know but "who" you know. Two weeks after this meeting Hugo received a phone call informing him that his applications had been approved.

Overjoyed he rushed to the police station, only to be told that he was allowed to keep two firearms only and he either had to send the rest back to Canada or sell them. Selling them was easier than sending them back. The police officer made a few phone calls and announced he had a buyer.

Within fifteen minutes Mr. Mes, the Minister of Human Development in the then Peoples' United Party (PUP) government arrived from the House of Assembly with his policeman son-in-law in tow. They wanted to take possession that day and demanded that the customs in Belize City release the guns that same afternoon. Therefore without further delay, Hugo was picked up in the government vehicle and they set off for Belize City. The minister bought two firearms and his son-in-law bought one and they were issued their licenses without any delay.

A few months later the son-in-law arrived at our house with a complaint. He said the minister was not happy because the handgun he had bought was misfiring. Hugo saw that the gun had not been cleaned and oiled. After he gave the policeman a short lesson in how to look after firearms, the last Hugo heard was that they were happy. They should have been since Christmas had come early for them.

In January 2007, the local newspapers in Belize reported that Mr. Mes had been charged with "grievous bodily harm and failing to provide a blood sample" after seriously injuring two village children. Not surprisingly, the magistrate ruled that there was not enough evidence to convict him. Getting off scot-free was the norm for the politically or other well connected classes in Belize.

An expat friend, who had lived in Belize for many years and knew about all the intrigue, told Hugo that the delay in receiving his gun license was due to the assistant commissioner holding out for a "blue-boy": bribery he was apparently notorious for. Paying an official a "blue-boy" to get results or service in a timely manner is known amongst the gringos as fast tracking. Fast tracking is not always a guarantee, but the financial loss one can sustain certainly is.

We only fast tracked once and successfully too. Hugo bought a 9mm pistol and I inherited the .22. We each had a Special Protection License, with the annual renewal gouging-fee of five hundred dollars for "foreigners" as opposed to seventy-five dollars for residents, falling due on one's birthday. Hugo's birthday was not due for two months and he had a choice of paying five hundred dollars then and again on his birthday, or leaving his new gun in police custody. He wanted to do neither, as behind the sunny face of Belize is a lot of violent crime. When Hugo hauled out all the cash to pay for my license, we could sense by the attending policeman's body language that a "blue-boy" would work. Sometimes one just had to hold one's nose and do what it takes.

Chapter 9

For where God built a church, there the Devil would also build a chapel.

Martin Luther (1483-1546)

Feeling we had found our feet in Belmopan, we were ready to join a church community. Belmopan Baptist Church congregation was a good mix of Creole, Hispanic and Maya, from various levels of education and prosperity, or lack thereof.

We were warmly welcomed. There were about ten of us gringos. Most were Americans who were in Belize on some or other Christian mission, medical or other humanitarian project in the villages close to the city. The church was a humble building with louvered windows and a low plasterboard ceiling with wobbly fans intermittently spaced.

I always aimed for a bench closest to the window to benefit from a breeze if there was one. On extremely hot days I would drape a wet hand towel around my neck, which had a cooling effect. If the fans were turned up too high the whirring noise made hearing untenable. That was not altogether a bad thing. A teenage band, mostly with drums and no piano, provided the music. The voices, singing with gusto, could sometimes be hard on the ears, especially with poor acoustics due to the low ceiling. There were certain ladies in front of whom one did not want to sit. Their enthusiastic singing bounced off the low ceiling with not-too-harmonious results. More than once during the singing time, I had to stand outside the church door to protect my hearing. However, who were we to complain or judge, knowing that God listened to the intent of the heart and not the quality of the voice.

Belmopan Baptist Church

We made our wishes to contribute to the church and community be known. Hugo joined in on the infrequent men's outings. The first memorable one was a fishing trip to Crooked Tree lagoon north of Belize City.

It is a large expanse of water not more than chest deep. Half of the group waded through, beating the water with sticks to drive the fish into a big net being held by the other half some distance away. When the two groups came together they hauled the catch into a dugout canoe they had dragged behind.

Fishing in Crooked Tree lagoon

In the catch was a three-foot crocodile that had become entangled in the net. Someone produced a previously unnoticed machete and immediately dispatched the unfortunate reptile. Crocodiles are supposed to be a protected species in Belize but no one seems to bother much with adhering to the laws.

On another occasion Hugo agreed to lead a group of men and teenage boys on a canoeing trip down a section of the Belize River. Hugo had been assured that all the Belizeans were familiar with canoeing and everyone had a life vest. There were about five canoes. Paul, a friend from the British High Commission, joined Hugo in his kayak. Because of the rains the river was flowing fairly swiftly. Within the first fifty yards, the first canoe capsized.

Fortunately it washed up against the riverbank where the occupants hung onto the thorny bamboo thickets that line the bank.

Belize River canoe trip

One of the boys didn't know how to swim, so several canoes went to the rescue. The canoe was righted, the water baled out and everyone set off again. It was then that Hugo discovered that most of the locals, including the pastor, had never actually been in a canoe before. While drifting down the river he had to bellow instructions on how to paddle. Thankfully the only loss that day was Paul's lunch and his camera.

To help strengthen the often-neglected bond between fathers and sons, Hugo helped organise a father-and-son bike rally that was to be a quarterly event. The rally would also include fatherless boys with mentors. With a logo especially designed by Hugo, with much excitement the first rally set off one Saturday. That first rally was also the last. We had begun to notice a general apathy and inertia amongst Belizeans, possibly the result of disconnected families; a dependency on aid from a multitude of outside sources; a sense of hopelessness, especially for the poor, brought on by wide-spread corruption.

Word spread amongst the Christian community that someone had a copy of the much-publicized movie *The Passion of the Christ*. As Belmopan didn't have a movie theatre it was to be shown in the local run down hotel. The movie was projected onto a wall that was not large enough to completely accommodate the picture and the dialogue. No one seemed to care.

Those who did not have a good view from their plastic chairs stood at the back of the room and the children sat on the floor in the front. There weren't any parents who thought of shielding their children from the meaningful suffering of Christ on the cross, the cornerstone of their faith. None of the children cried or appeared traumatised. They accepted suffering, hardships, life and death as part of the real cycle of life; quite a contrast to the need to sanitise life, as is so often the case in North America.

***

I had offered to be useful, but was quite unprepared for what was dropped in my lap. I was asked to oversee a lunchtime feeding program funded by a couple from the United States for the most needy of children who attended two primary schools close to the church. Apart from overseeing, I agreed to help serve lunches one or two mornings a week, thinking some of the mothers would also volunteer their services since their children were the recipients of this goodwill.

Besides Miss Paula the cook, a mild-mannered little Maya lady in her fifties who would receive a stipend, no one else was forthcoming. I found myself totally responsible for keeping this project going. Never having undertaken anything like this before, the first few weeks were like groping in the dark. I had no idea how to cater for the Belizean poor. The butane gas tank in the kitchen behind the church was empty. The facility itself would not have passed muster. There were no adequate cooking pots for the "stew chicken" and rice and beans and we did not know how many children to cater for. Before long, Miss Paula wanted to quit because the children were unruly and she felt she could not handle anything besides the cooking.

Slowly things started coming together. The schools submitted the names of thirty-seven needy children and only two mothers volunteered to help Miss Paula in the kitchen. Every Friday morning I would take the cook to the market for the necessary vegetables and to the Mennonite poultry shop to buy the week's supply of chicken and rice. I initially had a budget of US$1000 to feed 37 children five meals per week over a period of 10 weeks. I took encouragement from the miracle of Jesus feeding the multitude with five loaves and two fishes.

The children were supposed to contribute 25 cents per lunch. If they could not afford to, we did not insist. No doubt there were some children who were more needy than others; one could generally tell how hungry they were by the enthusiasm with which they ate their food. Many of the children arrived for lunch with cheap sweets in their pockets, which they bought from the candy vendor who regularly appeared outside the schools at lunch break. As there wasn't anyone else to take charge, I ended up supervising the children every day.

This feeding program did not fit the impression one gets when watching television commercials for the "Save the Children Fund" where one sees images of hungry little children on their best behaviour eagerly lining up for food. Many here were like little wild animals, throwing their chicken bones and rice and beans on the floor and everywhere else. On the last day of school I would take them cupcakes or a jelly pudding and fresh fruit juice as a treat. Afterwards I had to hose out the place as there was food thrown everywhere: there was even jelly clinging to the ceiling. Hosing down the room was a good thing as the doors had large gaps above the rough concrete floor and did not stay clean for long.

Miss Paula lived in close proximity to her two sons in Las Flores, an area against a hill on the outskirts of town. She had to walk to the church and home every day, which took her 45 minutes each way. In the heat and wearing flimsy shoes, it was not surprising that at her age she decided at the end of that school year not to return after the holidays. I was happy for the break too.

When school resumed in September, the American couple faithfully donated another cheque. The position of cook was taken over by Imogene, a large and strong Creole woman from the congregation. She was an able cook and organiser and had no qualms about controlling the children. This was a relief for me, as she did not need my regular daily presence.

It was enough that I had to see that the food was purchased weekly, keep control of the bank account and draw money every Friday. I was often the only gringo in the bank queue, sometimes waiting for over an hour on paydays. No one but I seemed to be in a hurry.

After the weekly money was drawn I would bump over the potholes on my way to pick Imogene up at her house. First we would go to the poultry shop to buy the next week's supply of a sack of rice and crate of chicken and then go to the market to buy dried beans, cabbage, carrots, onion and spices. The market stop was for Imogene the equivalent of going to the mall for gringo women. She shopped and socialised while I waited in the car chomping at the bit.

Eventually we came to an agreement to spend not more than one hour of socialising at the market and each time I synchronized our watches so that there would be no excuses. Belizeans do not regard punctuality as important. They joke about "Belize time" which is totally acceptable to most of them, even at official diplomatic functions. That always irritated the gringos.

To while away the hour, I started exploring the various makeshift second hand clothing tents. If one was familiar with type and quality of fabric and brand names there were some good bargains to be had. Hugo was kept in khaki shorts at six dollars a piece, as well as good brand name cotton Hawaii shirts, ideal for the climate and not the style favoured by the locals. To suit my own needs in the hot and humid climate, I sometimes converted a few quality items of clothing into tropical wear.

Imogene and I often compared our shopping successes. One day a real find was a bright yellow hand-rolled French silk scarf for two dollars. Not suitable of course for the tropics, but it certainly was for Canada. I was generally the only gringo scrounging amongst the clothes racks and it would certainly have been rather infra dig for the socially prominent ladies of Belmopan to be seen doing the same.

Imogene and I maintained a mutually deferential relationship. She called me "Siesta", as is common with the church folk, and I reciprocated. Kriol was her mother tongue and due to a lack of education, she could not speak English well, so our conversations remained perfunctory. She was self-taught to a certain degree and her potential intelligence enabled her to hide her lack well.

Like many, she had some strange ideas she shared with me; one being that if one washed one's hair too often, it would rot at the roots. I opined that there was a lot of ignorance in the general population, which most likely had more to do with educational deprivation rather than inherent stupidity although the latter was often a factor, especially evident when one was driving on the roads.

Food and butane costs were steadily increasing and it became more difficult over time to stretch the budget even though the American couple increased their donation. I was sensitive to Belizeans' dislike of foreigners being too authoritative over them; a hang over from colonial days I am sure. For that reason I did not keep as tight a reign over the program as I should have and allowed Imogene to keep the 25 cents from the children for incidental weekly food expenditure. Moreover, the pastor was too preoccupied with his other duties as school principal to give this program much attention.

One summer holiday Imogene went to visit relatives in Chicago for two months. There are about 55,000 Belizeans who live and work the United States, not including the ones who are there illegally. I had to wonder how her visit was possible on her stipend, even if relatives were subsidising her costs.

Due to the rising costs in food, when school resumed, I stressed the necessity of keeping an account of the daily "petty cash" as I called the 25 cents collected from the children, and that that would have to be handed over to me to add to the weekly expenditure. I was surprised at the amount of the weekly petty cash tally, and wondered if it had not partly subsidised the extended visit to the United States. When the summer holidays once more rolled around, I shared my concern with the pastor and expressed my belief that in all good conscience, under the present state of affairs, one could not ask for further donations from the American couple. He agreed, so after nearly five years, coinciding with the start of my bi-annual trips to Canada in 2009, my involvement with the feeding program came to an end.

***

Not only did our pastor lead a flock, he was also the principal of a high school he had established with the physical and financial assistance of certain churches and individuals in America. In our first year in Belmopan, we were invited to attend the school graduation ceremony. Fortunately the venue was held in a newly built air conditioned community centre. Once all the political and other guests were seated, the graduating students, dressed in their school uniforms over which they wore togas and mortar boards, began a slow march from the back of the hall to the beat of Elgar's "Pomp and Circumstance" march. One slow step forward and then a pause to about six beats of the march. It took a long time for them to reach their seats at the front of the hall.

After the usual valedictorian speech, came the main speaker who was a Belizean businessman. His aim was to encourage students that with determination and hard work, they could accomplish their dreams as he had done. It was rather touching to hear many students who had to face extreme odds caused by poverty, political corruption and favouritism, have the same dreams and aspirations as do students in the First World. We attended a few graduations ceremonies after that, not always in air-conditioned facilities. I always hoped that they would change to a quick march, but that never happened.

***

With time, we found it hard to sustain our church attendance. It seemed to us that more and more of what should have been preaching time, was given over to congregational input; sharing what came to some hearts and minds; someone bravely rendering a song; or the same little old lady, sadly lonely for attention, coming forward every week to hold her own mini sermon. It became a hardship to sit through it all, especially on a hot and humid day.

One Sunday, when the singing reached a painful crescendo, I turned to look at Hugo who stood with a scowl on his face and red earplugs sticking out of his ears. I knew there and then that that was the end of his church attendance. I persevered for a while longer. With the pastor juggling both jobs of shepherd and school principal, something had to suffer and the school demanded more of his attention. My decision to fall away too was well timed as I began my bi-annual trips to Canada.

***

Since Latin America is regarded as predominantly Roman Catholic, many people are surprised to learn that in Belize various Protestant denominations, including some sects, together make up about 25 percent of religious affiliations as opposed to 39 percent who are Catholic. Cults, other religions and those believing only in themselves make up the rest.

As most of the local church-going folk are not able to afford a vehicle, the churches each own their own bus; those ubiquitous recycled American school buses with a new identity painted on the side. Every Sunday one could see the buses bouncing over the potholes in the residential areas, picking up "sheep".

Many local churches are Belizean-planted, while some others are run by independent outside church organisations, mostly from the United States. While local pastors generally struggle to keep body and soul together for both themselves and their little churches, the wealthy outside organisations arrive with gifts of bicycles and other enticing items that attract congregants who have hopes of further handouts. It is not surprising that local pastors consider this an unfair method of sheep stealing.

There are many mission and other humanitarian groups that provide much needed and valuable services to various people of all ages. Amongst these are medical and dental volunteers who pay annual visits to regional hospitals and mission-run clinics to perform surgeries and other necessary clinical work that would be unavailable otherwise. There are instances where certain mission organisations make it possible for children to go to the United States to receive corrective surgery and life-saving treatments.

However, living in Belize has allowed us to have a different perspective on charity, mostly coming from North America. The negative aspects often come with "Short Term Missions". When one flies to Belize during school holiday time, one invariably will see a young "team" on the plane, all wearing similar colour T-shirts printed with a special trip logo. They brim with excitement over the exotic destination where they are going to bring "hope and relief" to the local people. Organising a team trip has become a million dollar industry. The teams hold fundraising activities and receive donations to finance their trips, often from well meaning but naive congregants.

Once in Belize, they set about with their "relief" work, which generally involves doing light repair or repainting work on a local church. The teams sometimes visit local children's shelters, where they supervise the children at play. At times some teams bring bible study or Sunday school material that the local church is unfamiliar with. The "hope" comes in the form of plays or skits held during a Sunday church service, which has no meaning for the local folk who are not familiar with the sophistication of it all or even the language.

After a week or ten days the teams return home with feelings of having gained missionary experience. What they have really gained is hopefully gratitude for their privileged lives, having observed how the less fortunate live. The local people continue their lives as before with no relief and no change, just waiting for the next team to do another upgrade to their facilities.

We watched many teams come and go. At the church we attended, the last team was unable to finish the paint job because the rain interfered and their time was up. My frequent suggestions to the pastor that the congregation join in and help the team was always received with little enthusiasm and no action. Gringo friends who were on a church committee in another town, suggested that their church needed to be repainted. A local committee member vetoed that idea, saying that they should, "Wait for a team to come."

As we had observed, many of these short-term teams lacked real understanding of the local culture and their needs, and their efforts did not alleviate poverty or bring a positive change to the lives of local people: instead, they added to the dependency already created by the dozens of foreign aid and non-governmental aid organisations. Their efforts would be far more worthwhile if they liaised with mission organisations that have already established meaningful work in the country.

By way of example, in his book *Toxic Charity*[8] Robert Lupton describes the response of short-term teams from America after Mitch (a category five hurricane in 1998) caused devastation to many parts of Central America. Teams, who went to Honduras to rebuild homes, spent about $30,000 per home, whereas local Hondurans could have rebuilt their homes for $3000 each. The amount another team spent to finance their trip to repaint an orphanage, could have created work for two local painters to do the job plus affording more teachers and new school uniforms for the students.

Chapter 10

Buy* (build*) a house in a foreign country and it seems, that anything which can go wrong, usually does.

Tahir Shah (b.1966)

Close to the time we were to start our house building adventure, John, a man from the local church offered to help Hugo do more bush clearing.

Ready for building

Understandably an opportunity is never missed to make a little extra money. We liked his suggestion of building a rectangular *palapa* to one side of where the house would stand. That would add a lovely tropical feature to our intended garden. Therefore, with two of his friends, each with a machete, they went into the bush for a day in a borrowed rattle-trap unlicensed lorry to cut poles for the structure and *huano* palm leaves for the roof thatch. They had to travel via the back roads to avoid police who most likely would have expected a bribe in return for not giving them a ticket for overloading and having an unlicensed vehicle. The next day Hugo made a few trips into the bush with the Nissan and the trailer to load up the palm leaves that had been bundled together with natural vines from the jungle.

Once all the material had been collected the upright poles and roof trusses were erected, using only a hammer, a handful of three-inch nails and a stepladder. The *huano* palm leaves used for making *palapa* and traditional Maya stick-house roofs, were intricately woven in an overlapping manner, which made them into a strong and rainproof thatch. We were pleased with the results and the men were generously paid.

John believed that his *palapa* building expertise would influence us to hire him as our building contractor but he was overly optimistic. At least now there was shelter for the cement bags and the night watchman, Ernesto. In his younger days he had been a bank guard, now he was a reserved, wiry older man, hired for us by the builders and came with a licensed rusty single shot sixteen-gauge shotgun. He made a make-shift bench with branches to go under the *palapa* and always wore long sleeves. Even so, I wondered how he could withstand all the nocturnal biting insects.

By the beginning of June we had temporary electrical and water connection and we were ready to proceed. Hugo was the contractor and would hold the purse strings. The subcontractors were the brothers Jose and Omar. Jose was a non-descript little man while Omar was striking with a mop of shoulder-length, curly dark hair that made him look more like one of Pancho Villa's bandits. We rather liked them. They calculated the needed building materials; which ones they considered the best buy and gave a lot of necessary advice. Hugo would do all the ordering and purchasing.

The building was a family affair and the brothers hired their relatives on an "as needed basis", paying them a daily wage according to their experience and ability. They piled into a well-used Toyota lorry every morning and travelled approximately fifty kilometres from their home village of Succotz to Belmopan.

The builders

Jose charged each person a small daily fee to cover the price of petrol. It was not more than they would have had to pay if they took the chicken bus and certainly much quicker.

At the end of each week, the brothers would present Hugo with a tally of the hours each one had worked so he could draw the wages. The brothers paid out the wages and were also responsible for making the Social Security Board deductions for unemployment benefits of sorts. The builders worked half days on Saturdays for which they received a full day's wage. We felt satisfied that we were paying them a more generous wage than was the local going rate.

The house would be built of cement block. The foundations were all dug by hand; not an easy feat in the hard clay and underlying marl soil.

The rebar for the foundation was bent and tied by hand in the shade of the trees. The brothers had a large block of hard wood, in which a few large nails were hammered and used as the rebar-bending tool. A cement mixer for the concrete foundation was rented from an individual in the nearby rough village of Roaring Creek.

Pouring the foundation

After the foundation had cured for a few days, three rows of blocks were built on top to elevate the house higher than the usual one row height, as a precaution against flooding. The space was filled in and levelled with soil and a hose was left running for two days to compact the soil. After that a slightly raised mesh of rebar was placed on top of the soil along with the necessary plumbing pipes.

The day arrived to pour the floor slab. A second cement mixer was rented and both were churning away full time. A row of builders hurried with wheelbarrows to rush the concrete from the mixers to pour the slab, while others stood by to level and smooth the poured concrete. It was all accomplished in one long day. Everyone was tired and satisfied with a job well done. It was definitely a far cry from the labour-saving cement mixer trucks used in the developed world.

Next came building the walls and the cement for this phase would be mixed by hand, on the ground. The bricks were ordered from a Korean who ran a brick factory. The delivery charge was two cents per brick. When they were unloaded, the bricks had to be counted, as it was common knowledge that often bricks would be missing. Either the driver had deliberately offloaded some somewhere else, or the factory had done some miscalculation. Whether the latter was deliberate or not, one never could be certain. Once we were two hundred bricks short so Hugo phoned the factory and received the excuse, "Oh yes, we not have enough, we send tomorrow."

A Mennonite company delivered the sand and gravel. A sawmill located at Central Farm, about thirty kilometres to the west, delivered the wood for the forms and roof trusses. There was a sawmill in Belmopan, but delivering wood was not part of their service. It remained a mystery to me how one could run a business of that nature without a delivery service. Not everyone owned a lorry. With so much political shenanigans and corruption, many things in Belize remained a mystery.

***

Besides building, there were also a few other events that drew our attention. One day we found three emaciated puppies of unknown parentage curled up on our building sand to keep warm. We discovered that they hung around a dilapidated wooden shack some yards up the road from our property. The shack seemed to be used by transients, who must have moved on and left the pups behind. We had no idea if and when anyone would return so we put out daily food and water. Before long only two pups appeared on the sand pile, and then none. We could smell their death in the adjacent bush. Noticing neglected pets would become a common and always disturbing sight.

However, the sand pile also brought life. Colourful lizards called Rainbow Ameiva, basilisks and geckos laid pockets of eggs in the warm sand and before long we had the first of many little creatures that would come to share our home and garden.

***

After the walls were built another important feature to be poured was the belt beam.

Pouring the belt beam

A cement mixer was again rented for this task. A wooden form was nailed onto the outside walls and an 18- inch deep reinforced concrete "cap" was poured on top of the walls. To make the roof hurricane proof, steel plates were embedded into the belt-beam to which the roof rafters were bolted. The joists and rafters were bolted together with angled steel plates, and the steel roof sheeting was attached with lag bolts.

The Mennonites of Spanish Lookout manufactured and pre-painted the steel roofing sheets, and our builders cut them to size with a hammer and a machete. We now felt confident that the house would hold together, should we be slammed by a hurricane.

Wanting a comfortable kitchen, with light cupboards so that there were no dark corners for cockroaches to hide in, Anne recommended a cabinetmaker. Jaime, who was a transplanted Colombian, had a kitchen cabinetry workshop and custom made whatever one required.

After a satisfactory consultation, and even giving him the standard measurements of heights and depths just to be on the safe side, he put his hand on my arm and said, "Tasha, you no worry. I meka you a bootiful kitchen."

Then with two cheek-to-cheek kisses that I soon learned was a common way of greeting amongst some locals and foreigners, even if they did not like you, I left feeling reassured that I would get what I wanted; lower drawers instead of shelves and all in a white duco finish.

With great excitement on my part, the day to install the cabinets arrived: the measurements were all wrong and I was fit to be tied. All the upper cabinets were way too high and I could not reach any of the shelves without standing on a footstool. The vertical row of smaller lower drawers could not open because they bumped into the stove. The bigger drawers were only three-quarters of the depth of the countertop, and the drawer runners were the flimsy kind used in bedroom dressers, and could not hold the weight of my crockery.

The installer-fellow, who had never seen a dishwasher before, wanted to cut the countertop at the wrong end to remedy the drawer-bumping problem. I had to stop him and give the instructions, after which he told me it was a good thing I was there to give him advise. After I created a lot of angry commotion, the cupboards were re-hung, the drawers were redone to the proper size and heavy-duty runners were installed.

***

As interesting as our retirement adventure was, it was not all sunshine and roses for me. My husband developed what I could only regard as Post Retirement Stress Disorder (PRSD). We seemed to have difficulty agreeing on many matters regarding the house and finishing. I was never sure from one day to the next if I would be waking up to Dr. Jekyll or Mr. Hyde. To keep the peace I acquiesced on many things. However, there is no such thing as peace at any price, and I needed to put some distance between all that was happening and myself.

In September 2004, nearly a year after arriving in the country, I flew to New Hampshire for two weeks of rest and recuperation with our son William. He waited patiently as I rushed around the supermarket, "oohing" and "aahing" at all the produce I had forgotten even existed and satisfying my visual senses that felt deprived after nearly a year in Belize.

As the building hardware stores in Belize did not offer much of a variety of small home fittings, I intended to fill my suitcases. At Home Depot I bought Kohler taps, white porcelain bathroom fittings to match the larger fixtures, and bedside lamps that switch on with a touch. One day William dropped me at the Mall of America on his way to work and picked me up on his way home.

Not even in Canada had I seen such a large mall, and being a great browser, I had no trouble passing the time. I popped into every store I passed and bought many items that I knew weren't available in Belize. After a visit that included a few good steak dinners and tasting some good wine, I felt fortified enough to return home and continue our adventure.

We had still not received those promised QRP identity cards. When I returned to Belize I was met with hostility from a female Creole immigration officer when I asked her not to put a visa stamp in my passport, as I had written proof from the Tourist Board of my acceptance in the QRP program. She said she was not interested in my piece of paper, and banged the visa stamp in my passport. That was the first time I had a sense of not being really welcome in the country, although our money was.

It was two and a half years later when we eventually received our QRP cards; only to find out then that contrary to the original information, the cards would have to be annually renewed for a fee. I was beginning to wonder if retirement was only a "win-win" situation for the government of Belize.

***

The brothers were good masons, but they were not jack-of-all-trades as they had led us to believe. The first snag we hit was in the plumbing department. They put the pipes coming out of the walls too high to match up with the traps behind the bathroom pedestal basins and the bidet. After much grumbling and blaming us for buying the wrong fixtures, they had to chisel the concrete away and re-position the pipes.

They also had no idea how to do the plumbing for a kitchen sink garburator and Hugo ended up completing that job.

The next snag came with painting the inside of the house. Hugo had drawn a diagram of all the interior walls, indicating the number and type of paint to use on each wall. They displayed great consternation that the walls were not going to be all in white, exclaiming, "*Wi noh du ih laka dat in Beleez!*" (We do not do it like that in Belize). That phrase was the most common one heard by anyone who has built a house in the country.

We encountered resistance in many areas, instead of them taking the opportunity to learn new ideas. I ended up doing much of the repainting, as the brothers seemed not to know the difference between latex and oil based paint and got the walls all mixed up, again blaming us.

The house was nearing completion, but we could not dismiss the night watchman, as the man who was supposed to have made the security bars was not forthcoming. He had been recommended by the brothers, and against his better judgment, Hugo had paid him the full amount up front. After much complaining, the security bar man arrived with the bars (still wet from the paint) and the rush job was evident in the big rough globs of solder iron left on the bars: they were ill fitting to boot. After much hissing on my part, the man returned to grind down the blobs. No one thought to remove the glass louvers while he did the grinding, and all the flying sparks burned rough little pit marks in the glass.

It became evident that the builders were getting tired of the job. Jose wanted more money than what was initially agreed and signed upon. When asked to justify the extra amount, the brothers came up with nonsensical reasons such as painting the fascia boards; filling in over the septic tank; the extra hours spent redoing their own plumbing mistakes in the bathrooms; having had to tile the kitchen and bathrooms, again saying, "*Wi noh du ih laka dat in Beleez*"*;* even though tiling was exactly what they were doing when we first went to see their work before we hired them.

Around this time someone from the Social Security Board came looking for Jose at the building site. He had not made any deductions from the workers' paycheques and owed them money. The extra money Jose wanted from us was the exact amount they owed Social Security.

We did not pay them all they wanted, so they left leaving many small jobs undone, which Hugo had to attend to.

We were happy to see the end of them, even though we were left to clean up the haze and grout residue left on the tile floors. After many unsuccessful attempts at scrubbing, someone told us to use muriatic acid, which is very corrosive. By then our furniture was in place and we had to move it out room by room to do the cleaning. Many broom-heads later the job was successfully done.

None of the books we had read could have prepared us for the actual reality of building in Belize, which turned out to be quite a frustrating experience for the most part. The main reasons were the cultural differences in taste, lack of efficiency and work ethic. As the contractor, Hugo had to be on the spot at all times to see that all things were correctly measured and in the right place and position. Our building experience was of course not unique, and we would later have a good laugh when comparing notes with others who had gone through the same experience. In spite of everything, we were for the most part satisfied with our new home and eager to move in.

Having said that, very dissatisfied was the recently arrived Haitian-American couple we met at the church. Since they had been scammed twice, they turned to us for advice because we had successfully completed our land and building deals. We recommended Arturo, who in spite of his name was a Chinese engineer and builder, who had worked on the controversial Chalillo Dam and had done a successful addition to our house.

They hired him to build a small hotel, which unbeknownst to any of us was financed by someone else. They were hired to run the project from building to eventually managing the "resort". Arturo did to them what was common in Belize (what our security bar welder had done to us); he spent some of the advanced money, ran short of cash and ended up unable to proceed according to plan.

The Haitians, especially the wife, made no attempt to hide her anger and demanded a meeting with all concerned on our stoop. Arturo was apologetic and offered collateral, which they refused. They wanted to hold Hugo and me responsible because we recommended him; they figured as gringos we had deep pockets. When they were not successful, her parting shot to me was that I was a "blue-eyed American". I had never heard that term before and was told that it was quite an insulting reference to a White person.

***

Danny, our young Belizean neighbour across the road, owned an unmarked bread van and offered to get a gang together to move our furniture from the rental house in mid-November of 2004. It had rained heavily the two preceding days, so moving turned out to be a muddy affair. We laid down strips of packing box cardboard from the van to the front door so that the fellows could traverse the mud with all our goods. Our round teak dining table top was merrily rolled without a care in the world across the by now dirty cardboard.

When the unloading was completed, we felt our possessions had sustained more bumps and scratches in the space of a few miles than all the way from Canada to Belmopan. Hugo's father always maintained that two moves equalled one fire.

Danny, who was of undeterminable cultural fusion, always gave us a friendly wave. He and his van were sometimes gone for days on end. He had money, but we could never figure out how driving the bread van afforded him his boat, cars and the additional houses he was in a never-ending process of building for his siblings on his five-acre property. The scuttlebutt was that he was a small time drug dealer.

I saw him up close in the bank line up one day. He was dripping with gold chains with dollar sign pendants around his neck, and gold finger rings to match. As long as we were on friendly neighbourly terms, who was I to care what he did. However, I did care about his garbage that sometimes piled up in his yard opposite our gate. The acrid smell that wafted through our window at night was the tell tale evidence of his nocturnal garbage burning. He acquiesced to my friendly request that he burn his garbage away from our gate. Danny moved on to more lucrative ventures and rented out his houses.

***

Being in a home we could once again call our own, and surrounded by our familiar earthly goods, did much to give us a sense of comfort and belonging. What really surprised us was the difference we noticed in the demeanour of our pets, especially our two cats. They perked up tremendously as soon as we had moved in. I am sure they sensed the difference between a rental and their new home.

The double arched screened in security door into the entrance gave them opportunity to view the outside world from the safety within, and the potted palms, where tiny little tree frogs took up residence, became their little hunting ground. They instinctively knew not to touch scorpions that had also come to reconnoitre for a new water source; their body language, portraying guarded curiosity, alerted us to go in for the kill with the can of Raid. They did not succumb as quickly as cockroaches did.

Our pre-occupation with building the house left us feeling somewhat socially isolated. Now that it was completed, we were able to invite new-found friends over. However, food would require greater thought and ingenuity, considering the limited ingredient choices and the humidity that melted some foods in a matter of minutes.

Our large back stoop would be the venue and we placed citronella oil lamps in strategic places in the evenings to keep the biting insects at bay. By this time we had met quite an assortment of expats, mostly American, British and some from various parts of Europe, many nomadic by nature.

Most of the people we befriended up till this point were scattered around the country, mainly down south in the Stann Creek and Sittee River areas. Therefore, visits were daytime affairs; no one wanted to endure the stress of driving on bad roads after dark or perhaps meeting up with a bandit or two, although the latter could happen at any time of the day.

I felt I was getting my life back, especially with my own electrical appliances hooked up again. William would be visiting over Christmas and bringing a few luxuries items like good coffee, chocolates and other goodies, and we expected fifteen people for our first New Year's lunch on the stoop of our new house.

There were still a hundred and one small jobs to be completed around the house. While visiting, William helped to pour concrete pathways and stepping circles around the house, and used his artistic skills by pressing big leaves into the wet cement, which left pretty patterns once the cement was dry. The driveway was built up with truckloads of marl and topped with gravel.

Hugo and I laboured over low concrete retaining walls to keep the gravel in place. We also spread gravel some distance beyond the pathways to keep the mud at a distance until the lawn grew. The mud on the German Shepherds' tails left feathery iron oxide stains on our clothes as they swished against us in a friendly greeting. We eventually had stained clothes kept strictly for garden wear.

Our yard had been fenced in with a six-foot chain link fence on three sides of the property and side fences up to the house to keep the dogs off the street. One day we heard the noise of a chainsaw in the front yard, only to find the electricity company cutting down a complete tree, which they considered one day might interfere with the electrical wiring, whereas removing a few limbs could have sufficed if the problem arose.

They did not consider it necessary to announce or even discuss their intentions. That incident spurred us on to secure our property with a front fence and a gate, which we would lock at night. With a ratchet winch and quite a few arguments along the way, Hugo and I got the job done. We attached an old-fashioned rubber bicycle horn to the gate that served as our doorbell. A few years later, in broad daylight, someone sneaked into the yard and stole the weekly housekeeper's bicycle from under our noses near the front door. The dogs gave the alarm, but by the time Hugo checked it was too late. He could just see the backend of the thief on the bicycle, disappearing down the road. After that, the gate was kept locked day and night.

We felt we were adjusting well to local life, perhaps Hugo a bit more than I was. We thought our acre would be a peaceful haven, and for the most part it was. However, there were some unexpected disturbances during many nights. Locals seemed to find it necessary to have gigantic base speakers blasting into the night from their numerous bars. The "thump-thump" from the distance reverberated through our concrete house almost like a jackhammer. Dogs, in distress and howling in the distance, were also another disturbing sound. To keep my sanity, I took to using earplugs at night.

***

There were no fibre-optic telephone lines, but at least we still had a cell phone. We also had cable television, with over one hundred channels in English, Spanish and Chinese. All the English-speaking channels were American, except for BBC World News; however, that channel disappeared when BBC discovered they were being pirated, or so we were told. If one was a soccer fan, there was no shortage of matches from anywhere to choose from.

The monthly cable fee was reasonable; justifiably so considering the cable was thin and strung along the trees. Consequently, our reception was poor and totally lost when it rained. We complained a lot to the cable company, whose policy it was to spend as little as possible on equipment. Eventually new management strung a thicker cable and the reception improved considerably.

We remained with our original Internet server, but soon our service became troublesome. We were not within a comfortable range of our server's antennae and experienced frequent disconnections, which caused them and us a lot of frustration.

Once again Hugo was told he would have to put down a $2000 deposit in order to get telephone connection. The Belizean woman who served him at the telephone company quietly suggest if he had a Belize driver's license he would be able to get a telephone connection for a deposit of only sixty dollars.

In order to get a local driver's license, he was required to get an eye check up at the Red Cross and a medical check up at the local hospital, where he was only asked if he was healthy. With those requirements completed, Hugo could now apply to Belize Telecommunication Limited for Internet and telephone service.

In spite of what the Belize Retirement Guide had stated, there was no fibre-optic telephone service in evidence, so we received Internet and phone service via radio antennae, which was common in many areas on the outskirts of town. The Internet was slow and erratic, but we had come to realise by now that the benefits of living in Belize we had read about were not necessarily the reality.

The cost of living was higher than the impression given in the Guide. Stating that Belize had the lowest cost of living in the Caribbean turned out to be somewhat misleading, since Belize is part of Central America, and had the highest rate in the region. Since there were no tempting shops, we spent money only on basic necessities. Even so, electricity, butane and petrol costs were high.

We realised that it was possible to live on US$450 only if one went native and lived in a stick-house and ate rice and beans and did without most amenities like hot water, a fridge, a freezer and other utilities that we took for granted. We were happy to downsize, but not to go backward.

Earlier in the year the government raised taxes on most items and the natives became restless. There were days of protests on the steps of the National Assembly in Belmopan, where the police, in brand new riot gear referred to by the press as "Gucci Gear", were pelted with stones and other missiles. We were impressed by the police restraint shown on local television, but a local man told me that the police would seek out the culprits after dark and "let them have it". There was a lot of big talk about shutting down the country and forcing the government to resign. The schools were the only places shut down for three weeks; water services were halted for a number of hours each day, when we would fall back on our bucket-and-jug system, and telephone and electrical services for a few consecutive nights. The latter caused me great irritation, for without fans, I found the humidity most uncomfortable.

I voiced my concern about people who might need emergency services during the shut down, but I was informed by Dewey that it would make little difference as the police never come when called and when the ambulance does eventually arrive, the patient would already be dead. The protests fizzled out eventually and things returned to normal. We felt sympathy for the little people because so many are caught in the crucible of government corruption and poverty.

Chapter 11

How fair is a garden amid the trials and passions of existence.

Benjamin Disraeli (1804-1881)

If you have a garden and a library, you have everything you need.

Cicero (106 BC-43 BC)

With the house completed and fences in place, we could devote most of our time to our main and most rewarding retirement occupation: tropical gardening. Having more enthusiasm than knowledge at first, it was survival of the fittest in our garden. However, since plants grow so vigorously in the tropics it was not difficult for most of them to thrive. Many farms and smallholdings have "living fences", as the wooden fence posts they use often spring to life and trees grow again. One only has to ensure that planting is done during the wet season, when the water is free and plentiful. Nature has ensured that indigenous plants can survive in the dry season.

During the wet season the rainfall usually occurs at night, sometimes in torrents, and sometimes accompanied by much thunder and lightening. During that time the humidity increases drastically and at certain times of the day the percentage is the same indoors and out.

We were ready for planting when the wet season started the first summer after the completion of our house. Lilly gave us a bucket full of plants and cuttings from her garden and from Anne and Noel (down the Hummingbird) we received sacks of lawn runners. No one seemed to know the type of grass, except that it spread fast and made a nice thick thatch that prevented any mud from seeping through. The back third of the property was fenced off and referred to as "Hugo's orchard". The name was not a true reflection, but rather where Hugo had the freedom to putter and plant whatever he wanted, without much interference from me.

Eventually we had variety of guava, most of which were devoured by flocks of parakeets and Mexican fruit flies. The avocado and papaya did not survive the waterlogged area where they were planted, but plantains, *pitaya* (dragon fruit) and passion fruit did well, the latter being at the top of our list of garden priorities. Some of our planted mango pips grew into substantial trees and one Cambodiana pip in particular produced delicious fruit, as did some of the grafted trees. Mango trees produce their fruit only when conditions have been favourable during blossom time. Often they rest one part of the tree during a season while another part will produce fruit.

We were surprised not to find passion fruit vines in Belize, which we thought would be in abundance. One day, by chance, I recognised a few passion fruit in the fresh produce section of a supermarket in Belize City. That was how we started our own production and passed on seedlings to other expats. The vines took about fifteen months to yield the first crop, after which they fruited about twice a year. Their skins were thick enough to be insect proof, as opposed to the thin wrinkly purple skinned variety one finds in Australia and South Africa. The pulp was also more tart than the southern variety. Thus, we had a constant supply of passion fruit juice concentrate I made and lots of pulp on hand in the freezer for making cheesecake and other culinary delights.

A bumper crop of passion fruit

After about five years, when the vines continued to flower but not produce fruit, we realised they had become sterile and we needed to introduce new DNA.

We purchased a rototiller at Spanish Lookout with high hopes of growing our own vegetables. The machine worked hard to dig up the clods of soil, which were too hard when it was dry and too muddy when it was wet. I became convinced that no genteel vegetable could possibly survive in that stuff, and it was small wonder that the robust carrots and potatoes procured at the local market looked the way they did. I gained a small modicum of compassion for the local farmers. Nevertheless, we planted carrots, Swiss chard, some watermelons and gem squashes. Like the guavas, it was a losing battle. One could not spray enough insecticide to overcome the entire insect onslaught, which started even below soil in some instances. The effort was too much for us and we conceded defeat and would leave the hard work to the local farmers.

Palms were next on our list of plant priorities.

Front entrance

Steve suggested we visit Lou, an expat American palm farmer and seed exporter, who lived near the end of Young Gal Road. He had an interesting garden next to the Belize River, where we discovered the wonderful world of palms. He was quite generous in supplying us with rare heliconias, delicious monsters and other exotic tropical plants; we eventually acquired twenty-three different species of palms from him. I thought he was a little on the anti-social side, especially towards women (there were a few of those around in Belize); however, he seemed to enjoy Hugo visiting him on his own turf.

We received a load of hibiscus cuttings from Paul of-the-golf-course, and planted them in a criss-cross manner along the side fences where they eventually formed a hedge. He also gave us many coconut palms, which with time provided us with our own supply of fresh coconut for culinary purposes and also enjoyed by the parrots. The jungle provided us with various bromeliads and a variety of philodendrons, which before long rooted onto some of our trees.

One day Dick and Peggy arrived with about seven different kinds of wild orchids. That introduction was the start of Hugo's fascination with them. Belize has over three hundred species of wild orchids, of which about 80 percent are epiphytic. They come in a variety of interesting colours, shapes and sizes from large to miniature. The black orchid is the national flower of Belize and it is illegal to harvest any orchid in the bush. However, many orange orchards are loaded with a variety of wild orchids, including the black one, so it is possible to acquire them. That does not mean of course that orange orchards remained the only source.

Over time Hugo had collected 47 species that he could identify and about a hundred and thirty plants in all. His orchids were all attached to the trees in the garden, numbered and catalogued and they never ceased to fascinate us. With time some propagated themselves and some died off. They were specifically pollinated by pretty and rarely seen orchid bees, which have a complicated relationship with the plants. Phalaenopsis orchids, the ones one usually sees in florists in North America, are cultivated in Belize by the Taiwanese. They were the gift of choice for birthdays or anniversaries, so I had a small collection of those, hanging under the Thunbergia-covered pergola at the front door.

We planted two Poinciana saplings given to us by Dick, which shot up to about twenty feet in three years and produced beautiful red blossoms in the spring. We helped ourselves to a few pink frangipani (plumeria) cuttings from a very old tree growing next to the Western Highway, between Belmopan and San Ignacio. A rarely seen cup-of-gold vine cutting I acquired after asking a stranger who had one growing in her front yard in another village.

One cannot live in the tropics and not have bougainvilleas. From Audrey we received cuttings of a prolific red one that spread quickly and grew into an archway over the gate and along the front fence. We also had salmon coloured ones; one in particular that grew over the roof of the large aviary. Eventually we also had a collection of gingers, various species of heliconias, ornamental bananas and other flowering vines, and over all, pleased with what we had created and the compliments we received.

***

Hugo had selectively cleared the trees on the property, leaving the ones that he knew would provide food for wild birds. With the bush cleared away, the trees had more access to sunshine and grew with renewed vigour. A few banana saplings that had been dormant underneath the tangled undergrowth sprang to life. We hung up hummingbird feeders, which hummers shared with a variety of colourful orioles. We rigged up an "avian café" on a tree and we stocked it daily with bananas and other fresh fruit, all within sight of the stoop and the kitchen window.

The stoop from the back garden

The fruit and water from a cement bath at each garden tap attracted a large variety of interesting and beautiful birds; amongst them were different coloured tanagers, catbirds, saltaters, tropical mockingbirds, Yucatan woodpeckers, kiskadees and social flycatchers, small toucans called aracaris, mot-mots and varieties of warblers, tropical wrens and euphonias.

We also had squirrels and basilisks that came to eat the fruit. We were hoping parrots would drop in, but they kept their distance. In the springtime the bird activity increased as they built nests in many different places in the garden and taught their young how to feed themselves. One kiskadee pair built their nest in the same tree year after year. We could always tell when the chicks flew the nest, as they screeched from the surrounding trees as their parents continued to cater to their hungry demands and hover over them for another few weeks.

On a few occasions hummingbirds built nests close to the stoop, where we had a good view of the whole process from eggs to fledglings flying the nest. Sadly many nestlings did not always have a happy ending. Predation of nests by snakes, lizards and other unknowns, was high.

***

Birds were not the only interesting creatures in our garden. Even though not as beautiful, the burning, biting and stinging varieties were equally fascinating! Fire ant nests occurred mainly in the wet season. They were hidden underneath the soil or gravel and only became evident when their vicious little bites on our feet and ankles caused us to jump and run for the Benadryl itch-relief cream, an essential when living in the tropics. There were black army ants too, that stung if one intruded on their long columns as they marched through our property in their quest for new nesting sites.

On their marches they would explore up and down anything that was in their path as they sought for food along their way; the insects on their route fled from the approaching danger. We could tell the progression of the army by the mobs of insect eating birds keeping pace as they fed on the dislodged and fleeing insects.

Groove-billed anis were particularly interesting birds. They were pitch black with beaks that looked like Roman noses and lived in small family groups. As soon as the lawnmower started up, they would appear from nowhere and run alongside the mower, sometimes inches from the blade, to catch the fleeing insects.

Occasionally the ants wanted to march through our house, but I diverted them with a can of Raid. I did not want to take the chance of them deciding to take up permanent residence somewhere in the ceiling of our house.

Leaf cutter ants, known locally as “wee-wee ants”, would forage night or day and carry the leaves like parasols to their underground nests. We would occasionally find a tree in the morning, stripped of all its leaves overnight. We also had to be aware of worms or caterpillars, whose hairs could sting as painfully as a scorpion if one rubbed against them. One particularly unpleasant one was a "saddle-back" caterpillar that was well camouflaged on palm fronds. Tarantulas also inhabited the garden, but they seemed rather shy and most likely saw us more than we saw them.

***

There was the potential for more serious dangers lurking in the garden by way of mosquitoes. We never encountered anyone who had contracted malaria, but dengue fever, a virus transmitted by Aedes mosquitoes, was always a possibility in the rainy season. One could identify the culprits by their white "feet", if that is what one can call the tips their legs. They generally bit one in the daytime, as opposed to the Anopheles mosquitoes whose legs are all black and who came to life after dark. Dengue fever, especially the hemorrhagic kind, can be serious. I experienced dengue fever twice, both times in June when the dengue threat is high, and fortunately only mild bouts. All I had to endure was an unpleasant fever with high volumes of sweating for four days.

A more horrific (in my opinion) and actually a marvel of nature that I fortunately escaped, but which Hugo was victim of on three different occasions was the *dermatobia hominis*, more commonly known as a botfly, or beef worm in Belize.

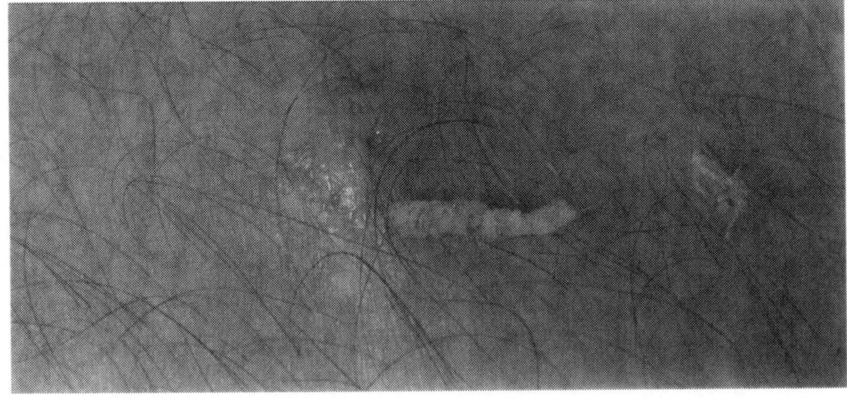

A beef worm from Hugo's leg

In this event the female botfly catches a mosquito in mid-air and attaches up to thirty of her eggs under the mosquito's wings. As soon as the mosquito lands on a warm mammal to feed, the body's heat causes the eggs to hatch immediately. The hatched larvae burrow into the skin where the mosquito had bitten and from there it grows into a fat, round maggot. One only becomes aware of this invasion when itchy bumps, like small boils, appear. If left untreated, painful pustules that secrete fluids will result.

For unsuspecting tourists returning home, and their medical doctors who are ignorant of the more unpleasant aspects of tropical travel, the evidentiary outcome of the pustules can create quite a sensation. When mature, the larvae will release their anchoring spikes and make their appearance through their breathing hole in the skin. They drop off and pupate in the ground, from which the new fly emerges and the cycle is repeated.

In reality they are easy to get rid of. One only needs to block the breathing hole, which then causes the maggot or larvae to suffocate. The cure: duct tape, Vaseline or anything that seals will work. If still small, one's own body will resorb the maggot after it dies; if not, one has to carefully squeeze the bump as one would a pimple and pull the dead maggot out with tweezers. Victims soon recognise the painful creepy feeling of the bump. Hugo used a drop of Ivermectin that works equally well on humans and animals. It was known that the indigenous people used chewed up raw tobacco placed over the bite, with a banana leave wrapped over to keep the wad in place. The nicotine, which was one of the earliest known insecticides, killed the maggot. It is advisable to wear a hat in the jungle, as one's head seems to be a common place to get a beef worm.

***

Now that we were settled and had enough garden space, we acquired a third German Shepherd; a cute little male puppy who had no ticks or fleas and who we named Inja, meaning dog in Zulu. He was well bred by Julian, who grew up in Belize with an expat South African father and an American mother. With his wife Olda, a Hispanic Belizean, they own a nice gift shop and small accommodation called Orange Gallery between Belmopan and San Ignacio.

Chapter 12

Wishing to be friends is quick work,
but friendship is a slow ripening fruit.

Aristotle (384 BC-322 BC)

When one makes a major move as we had done, there is always some sense of loss of friends and family one had to leave behind. I had felt that deeply when we emigrated from South Africa to Canada in 1972. However, this time we had not moved that far away so we looked forward to frequent visits from those near and dear to us. Our son William visited in the summer of 2005, followed at the end of the same year by Alice and Larry, our first friends from Canada to visit. We had a loosely held itinerary of places of interest to show all our visitors.

Always at the top of our list was Xunantunich, a small Maya ruin beyond San Ignacio, from whose top one had a far-reaching view into Guatemala.

The road west allowed visitors to see the variableness of Belize scenery, including the town of San Ignacio and the Mopan River which had to be crossed to reach the ruin. The river crossing was a unique experience for those unused to lack of progress, as one had to drive onto a pont that could hold three cars at most, which was then propelled to the opposite bank by a hand-winch.

Xunantunich

Crossing the Mopan River

Very often one could see three-foot long iguanas on the far bank, where they felt safer from the locals who like to catch and eat them, referring to them as "bamboo chicken".

Cahal Pech was another small ruin in the environs of San Ignacio worth visiting, were one could still see active archaeological excavations. Then we would stop for lunch at a little hole-in-the-wall restaurant run by Peter, an expat from Rhodesia (now renamed Zimbabwe) where the menu had tasty choices different from the usual fare of rice and beans. Our favourite was curry, made with lamb Peter raised on his own farm.

Next on our list was the Belize Zoo, almost halfway between Belmopan and Belize City. Following the gravel paths one has a sense of being in the bush as the mesh enclosures and fences are hardly noticeable: wearing a hat and well lathered with bug spray was a necessity. Going south down the Hummingbird Highway was always on the agenda as our visitors could see a real tropical jungle, the coastal areas and many interesting places along the way.

The Mennonite communities were other places of interest. They are self-sufficient communities and vary from progressive to traditional: each community lives according to the dictates laid down by their ruling elders. Spanish Lookout is a large progressive community, roughly thirty miles west of Belmopan. With their well-built and maintained roads, their neat and clean countryside and homes reminiscent of the 1950s (homes on the Canadian prairies from whence they originally came) one almost has a feeling of being in a different country.

Spanish Lookout provides most of one's hardware, building and mechanical needs, as well as all poultry, eggs and dairy. The various Mennonite communities throughout the country grow beans, corn, rice and watermelons. The Mennonites literally feed Belize.

We preferred using full cream shelf life cartoned milk produced in Holland rather than Mennonite milk. When their milk was delivered to the local supermarkets, one could often see the crates standing in the hot sun with the milk surely turning sour, while the stores took their time to pack the milk into their refrigerators. In addition, we could never find out how well their milk was pasteurised; being from the veterinary profession, we were naturally cautious as there was no TB or brucellosis testing done in the country at that time.

Traditional Mennonite selling watermelons

Eventually the government banned full cream shelf-life milk in order to give the Mennonites that monopoly. We then resorted to using full cream powdered milk produced in Holland and sometimes I was able to buy small cartons of fresh milk from America.

In the progressive Mennonite communities everything was modern, but the women seemed mostly pale and wan and they all looked rather similar in their same-styled dresses and little kerchiefs pinned to their hair. Apparently the women are not allowed to marry outside their community, but a man can if he brings his woman back into the community.

The traditional communities shun anything modern and use a horse and buggy for transport, even though their buggies have modern rubber tires. The women wear long dark dresses and peaked bonnets, reminiscent of the Boer women during the Great Trek in South Africa of old, and the men wear long black pants with braces, beards and straw hats. The children, even babies, all looked like clones of their parents. Just the thought of all those clothes completely covering one's body in the tropical heat and humidity made me want to itch. I wondered what sort of rashes and eczemas lay hidden underneath.

Of further interest was Springfield, a small and traditional Mennonite community just south of Belmopan. They were good gardeners and sold fruit trees and vegetables when the weather was not too severely hot, and eventually even added romaine lettuce to their selection.

They don't have any electricity and they literally use horsepower. Eight horses walking in a circle, power a sawmill. That power is transferred via an old lorry differential and a series of pulleys to a huge band saw. At another location a pair of horses, in a similar manner, power a chop saw for cutting firewood.

A Mennonite "logging truck"

Those same two horses also power a compressor for a self-taught dentist's drill. He makes his own dental plates, the teeth for which he orders from Guatemala. The poorer Mayas are mostly his clients, and it was quite easy to tell whom. Our gardener's mother was one of them.

Often a young man from Springfield came to our house to ask Hugo's advice on treating their horses or dogs that had specific health problems. He was pleased when Hugo presented him with a copy of *The Merck Veterinary Manual* that Hugo's diplomatic friend Jack had ordered.

Sometimes the traditional Mennonites stood at the roadside "praying" for a ride. They would never put out a thumb, but one could see the anticipation in their eyes. Once we picked up a well-spoken young man and I asked where they drew the line between what was modern and what was not, since they did not own cars but did not mind riding in them. He said they themselves were not quite sure. He said he was free to own a car but he would then have to leave his community. Whether traditional or modern in their beliefs, the Mennonite community is, without a doubt, a great asset to Belize.

***

John and Wendy decided on the spur of the moment in January 2006 to visit us after attending a conference in Texas, only a short two-hour flight away. Being ex-South Africans, John was always in search of sunshine. During their visit we experienced an unseasonal tropical storm, which kept us housebound for days. The yard was waterlogged and the toilets had difficulty flushing. Even though there were enough clear days to visit our prioritised sites, it was not the holiday John had anticipated and I could never persuade them to visit again. Apart from the weather, John was quite righteously vocal about all the garbage littering the countryside, contrary to the impression of an ecologically pristine rainforest one gets while perusing tourist brochures.

Two thousand and six was a busy year for visitors. In March, Hugo's 89 year-old mother detoured her return to South Africa from Canada to visit us. She was still sprightly enough to want to see all the sights and even made it to the top of Xunantunich, which rises 130 feet above the jungle floor. Esther, a friend from my heyday in South Africa, followed her a few months later.

A casual acquaintance of William's visited us from Canada during the summer. By now I was getting tired of all the trips to the top of Xunantunich and decided to remain in the car to read a book. The parking lot was a gravel patch, surrounded by grass and trees. Some Belize Defence Force soldiers, there to provide protection against bandits who occasionally robbed visitors, were lazing against a few picnic benches on one side. On the other side was an office of sorts with a uniformed individual sitting behind a glass window. Near the entrance to the parking lot, a factotum was cutting grass with a weed eater. All of a sudden a noise like a gunshot rang out behind me. In a split second the rear window of the Nissan started shattering into a thousand pieces.

My first thought was that a coconut had hit the window, and at the same time I realised there were no coconut palms overhead. By this time I was standing outside the vehicle but no one else in the vicinity seemed perturbed by the loud bang. My gaze caught sight of the man with the weed eater and I knew instantly what had happened. Knowing that as a woman in Belize I was not likely to elicit any respectful response from the men on duty, I waited for Hugo to return.

The man-behind-the-glass-window said he would have to find the gravel stone in the car to believe my assertions, so he came to look and there it was. It took another two hours for him "not" to be able to find a relevant form to fill in and then he told us we needed to be at the Benque Viejo police station at eight o'clock the following morning, about 35 miles from Belmopan.

The next morning we arrived on time and were ushered into a policeman's office. After silently looking at a report he said, "*Dis is a diffikolt kase to krak*" (This is a difficult case to crack). Expecting some stonewalling, and knowing my husband disliked confrontation, I jumped in and said that there was no difficulty as the gravel evidence was there for all to see and that all we expected was some financial contribution towards the window repair. We were dismissed and told the report would arrive in Belmopan in about two weeks.

However, word obviously travelled faster than the dispatch and a few days later two men from the Department of Archaeology stopped by the house to check the damage. One of them told Hugo to come by his office the next day, but when Hugo did so, the man was out. He was out every time Hugo went back until Hugo got tired and never went back again. That was the Belize bureaucracy's standard modus operandi, a common passive aggressive avoidance technique that one frequently encountered. It always brought to mind Rudyard Kipling's *The Story of West and East*, some of which I apologetically changed to, "For the gringo riles and the Belizean smiles, and he wears the gringo down", even if it was only for my own inner satisfaction.

After riding around for quite some time with plastic duct-taped over the back window, we had no other recourse but to have the window replaced at the Belize City Nissan dealer at four times the cost it would have been in Canada. While waiting at the dealership, we saw a brand new top-of-the-line leather upholstered Nissan with all the bells and whistles standing ready with a gigantic ribbon and bow tied around it. When I commented that someone was a lucky new owner, we were told it was for a government minister, whose driver arrived just then and drove it away.

Our sons William and Hugo Jr., and daughter-in-law Alison came for Christmas in 2006 and we were happy to have all the family together again. After Christmas we hired a young expat American girl to pet sit while we took a trip into Guatemala to visit Flores and Tikal.

We frequently heard stories of bandits holding up cars on remote roads, so I felt a little apprehensive. Belizeans blamed the Guatemalans for the banditry, and they in turn blamed the Belizeans. The first thirty miles of road across the border was unpaved and rough, which increased my fear of Hugo not being able to make a get-a-way if bandits should confront us. Fortunately, we encountered none and the only discomfort suffered was our three young adults who had to endure bumpy conditions due to their father's lead foot on the accelerator while they sat squashed in the back seat.

With Hugo Jr., Alison and William at Flores market

We stayed in the quaint little town of Flores and explored the many shops where one can buy pretty weavings the Guatemalans are known for. Many Guatemalan women also travel to San Pedro in Belize to sell their wares to tourists there. We enjoyed eating at interesting little restaurants with varied menus that did not include rice and beans.

We spent a long and interesting day at Tikal, the ruins of one of the most famous Maya cities which covers a vast area. Like all archaeological sites, it has an interesting history, including the act of human sacrifice, which is evident on their murals.

Tikal has a connection with Belize. In 553 "Lord Water" ascended the throne of Caracol, a Maya kingdom in South Western Belize, and he later conquered and sacrificed Tikal's king "Double Bird", and ruled Tikal and other Petén Kingdoms with an iron fist until the late seventh century. Guatemala has for years laid claim to Belize and that historical incident could very well have motivated lasting revenge that precipitated this latter day claim.

Belize has quite a number of important Maya ruins dotted around the country, which I was not keen on visiting. My belief was that once I had seen one, I had seen them all, and already I had seen three.

***

After only three years, our *palapa* showed signs of premature decay. The upright posts were infested with wood lice and the thatch was rotting. We found out that the tiny termites attacked because the wood for the posts was not the resistant wood that John had said it was, and the thatch could have lasted for years if the *palapa* had been built in the sun instead of in the shade. Early in 2007 when Walter and Sandra from Canada visited us, the two men set about dismantling and then burning the whole structure. We were very disappointed at the loss of this garden feature and felt we had been "Belized".

Larry and Alice continued their annual visits and we enjoyed many interesting excursions together. One jaunt was down south to a shrimp farm owned by a man from Louisiana (or thereabout) who was not allowed to sell his shrimp in Belize. Miss Emma, originally from the Philippines, was his housekeeper and excellent cook who provided all the meals for the farm workers and visitors alike.

Mr. Howard was generous towards us and after visiting we always came home with a cooler filled to the brim with freshly caught shrimp. By his own admission he was not popular with the locals as he was vocal about their general laziness, dishonesty and primitive custom of putting their toilet paper in a wastebasket instead of flushing it down the toilet.

Once home, it was Larry's job to sit on the stoop and behead the shrimp and readied to be packed into the freezer. Larry was not one to sit around the house: he and Hugo were always on the go. They kayaked the Belize River and some coastal lagoons down south, and the Caves Branch River, which flows through a series of long caves. They spent a few nights in the Toledo District and took a day trip to Tikal. Alice and I stayed at home, knowing how uncomfortable it was to be backseat passengers when Hugo was driving and Larry sitting in the front. We took our turn by choosing to go to the cays.

Alice in the background of a Rhyncholaelia digbyana orchid

On one trip to San Pedro, we were the sole customers at a beachside restaurant and therefore had an opportunity to chat to the young Creole waitress. In answer to our questions, she informed us that she had left her children in the care of a sister in Belmopan so she could earn more money in the tourist industry. She had a boyfriend who was the father of her third child, but she chose not to marry him because he cheated on her.

We were rather shocked to hear from her that single women from the United States came to the cay on occasion, in search of Belizean men with whom they could have sexual encounters: maybe we were just too old fashioned.

On a visit to Cay Caulker we shared the expense of the one good hotel on the cay. I did not want to risk a previous experience I had with another friend Linda. We stayed in a quaint looking hotel on the beach where I endured an excruciatingly painful night; the bed's base was made of concrete with an inferior mattress plunked on top, quite a common "furnishing" in Belize. To sleep on those concrete boxes one needed to be anaesthetized, which could well be the case with many younger tourists, since Cay Caulker is popular with back packers and "ganja" is the local smoke of choice and quite easily obtainable at a popular hangout called "Lazy Lizard" situated near "The Split": the name of the narrow channel that cuts the cay in half.

Chapter 13

We can judge the heart of a man by his treatment of animals.

Immanuel Kant (1724-1804)

Around the time of moving into our new home in November 2004, Lilly and Dewey put their efforts into starting a much needed Humane Society in Belmopan. It was natural for us to become involved in this undertaking. Apart from the gringos, the first committee included Belizeans concerned with animal welfare and this was a good thing. Knowing how much Belizeans resist foreigners telling them how to do things, these committee members could hold up front positions. One of the gringos, who also spearheaded this society, was a member of the British High Commission, through whom we came to meet other Brits. Expats were invited to go to the "Pig and Parrot", a British clubhouse where they could socialise over a glass of libation on a Wednesday evening.

Sometimes a few "big frogs" from the local pond made their appearance. I recall an occasion when one of them was holding court, seated on a high barstool, with all his sycophants hanging on his every word. He was puffing on a fat cigar and flicking the ash onto the clubhouse floor. There weren't any ashtrays in sight, which should have indicated to him that it was a no smoking area. That compelled me to offer him an aluminium pie plate to use instead.

With time, membership to the club was restricted to foreign office staff and invited guests. We appreciated being on the list; admission to the club added a pleasant dimension to our social life and expanded our circle of acquaintances to include more people of a similar feather. I looked longingly at the swimming pool next to the clubhouse, but alas, it was for foreign office staff only.

One of the first undertakings by Dewey and the up front members was to meet with the City Council to offer help and suggestions to create a more humane method of dog control other than the strychnine poisoning that was in use. It was disappointing, but not surprising, that the members' efforts were refused, considering the general attitude of resistance to change and the lack of compassion: attitudes I found hard to come to terms with.

Realising that educating people would be the best hope of improving life for animals in Belize, which certainly is not a paradise for them, Lilly and Dewey organised the first Pet Fair in Belmopan. People could bring their dogs and along with the fun activities, receive free health check ups, advice and rabies vaccinations. The dogs in Belmopan were in the same pathetic state of neglect as we had witnessed in Belize City. Many spent their lives attached to chains and were let loose at night or on garbage collection days so that they could scrounge in the bins for food.

Of course the problem of neglected dogs was greatly exacerbated by the fact that neutering pets was not on most Belizeans list of priorities. Neutering costs were beyond the reach of many folk, and added to this was a male dominated macho society where immature men equated their dog's masculinity with their own. Confirming my opinion, a prominent Belizean woman told me that Belizean men would never consider using a condom. Moreover, as women are the marginalised members of Belize society and generally bear many children, it was highly unlikely that they would even consider spaying a female dog.

At the Pet Fair Hugo worked alongside a local veterinarian whom he met for the first time. He later asked Hugo if he would work with him in his clinic, but Hugo, having seen his clinic, declined. He could not see himself working in such an unhygienic and ill-equipped establishment especially as his own hospital standard had been graded in the top 10 percent by the American Animal Hospital Association. We were shocked by the poor standard of veterinary practice. It seemed to be perpetuated by poorly educated veterinarians who were not held to any minimum standards, nor was there any requirement for continuing education. Added to this was a population who in most cases could not afford veterinary services.

Before long Hugo received many calls for help and advice regarding sick pets from both local and gringo pet owners. He helped as much as he was able from a retirement position without a clinic. Shortly after our arrival in the country, he had written to the local veterinary association to introduce himself, as is customary where we came from, and offered his volunteer services if needed. His letter was never acknowledged. Neither was the letter he had written to the zoo offering them his volunteer services. We felt it was unfortunate that they were reluctant to avail themselves of expertise people were willing to share.

With time Lilly herself became the driving force of the Humane Society, with Dewey as her Man Friday. When possible, she arranges free spay and neuter clinics by visiting volunteer veterinarians from the United States, engages the services of a local veterinarian for the same purposes, and through education and fundraising, subsidises the cost for locals who want their dogs neutered but cannot afford to pay.

They managed to get Belmopan City Council to donate a piece of property with the hope of eventually building a shelter and clinic, and she still tries hard to rehabilitate neglected strays and place them for adoption. I could only admire her ongoing dedication to what would have caused me too much grief to deal with on a daily basis. For her continuing efforts, I always felt Lilly deserves the "Nobel Pet Prize"[9].

***

Nikki and Jerry, a British expat couple who settled in Belize around the same time as we did, bought property near the rough village of Roaring Creek and started rescuing parrots and other birds. With the sanction of Belize Forestry Department, which is responsible for fauna and flora, they eventually turned it into "Belize Bird Rescue". The local pet trade and human encroachment has put many parrot species on the endangered list.

***

We had been in our new house for about a year when we acquired our first parrot. It happened quite by chance. Hugo was busy in the front garden when a local man cycled by, dangling a small bell-shaped cage with a Red Lored parrot in it. Hugo called out to him and asked where he was going with that bird, to which he replied, *"I di go sell mi buddy"* (I go sell my buddy). Hugo offered to pay him his asking price of forty dollars and so Bonita became our first "rescue".

She was about two years old and had lived all her captive life in the confines of that small cage. We converted the cats' travel crate into a temporary birdcage while Hugo set about building a substantial cage that could stand on our stoop. Bonita's muscles were so unfit that she could hardly move from one perch to the next and she was ravenously hungry since she had been raised mostly on tortillas. It took a few weeks for her to regain her muscle tone, to become well nourished and used to me handling her. The sounds she emulated were similar to those of a smoker's cough, cocks crowing, a small child calling its mother and a baby crying. Our housekeeper identified the language as that of a Maya family. Bonita came with clipped wings and therefore could not fly away, so she and I had many siestas together in a hammock.

The following year a Taiwanese acquaintance brought us a Red Lored chick that had ostensibly been rescued from a nest in a felled tree. Having raised a parrot in our South African years, this was not a completely new experience. She was not yet old enough to feed on her own, so I quickly learned how to make "parrot pablum" and stimulate her feeding responses. She thrived and soon became good company for Bonita. We named her Nyoni, meaning "bird" in Zulu.

In 2007 we also adopted an adult parrot that had been rescued after being hit over the head and blinded in one eye in a capture attempt by a local. Hugo had to make a second cage to house Frankie, who was a traumatised bird and never made a sound or took food from our fingers for the first year we had her.

Being in a small community, it did not take long for others to know one's business. Somehow word got around that we were gringos to make a few *dallas* from: local kids frequently arrived at our gate hoping to sell us the fruits of their nest robbing. We turned many little animals away but I never had the heart to refuse a parrot, knowing the fate that awaited these intelligent birds if we did.

Paying the kids of course created a vicious cycle. I knew I would not be stopping their trade, and condemning a parrot for the sake of a principle was not a choice I could make. In 2008 a local boy brought us Billy, who was a year old and the same age as Nyoni. By this time we had three home built cages and we realised it was necessary to start building proper aviaries where the birds had space to fly. Clipping wings and depriving a bird of that natural ability (an acceptable practise in the parrot pet trade) is similar to disabling a human to walk.

William came to visit again that Christmas, loaded with corner pipe connectors and set about building the first flight aviary. Half inch galvanized mesh was imbedded into a concrete floor, which assured that snakes and other undesirable creatures would be kept out. Hugo, with the aid of our gardener, added two more aviaries. About this time with Nyoni being one and a half years old, Bonita turned aggressive towards me with whom she had a love affair till now. We realised that Bonita had reached sexual maturity, and was in fact a male; henceforth, known as Bonito who transferred his affection from me to his new love, Nyoni.

Observing their behaviour after maturity was basically the only way we could tell what sex the parrots were.

In May 2008, the following year, just after breeding season, five chicks came our way from two different sources. Some of them were in poor condition, but with good nutrition they soon regained their health. Initially they were in a nursery cage and when feeding time arrived, they would perch in a row with their mouths open and wait their turn for the pipette tube filled with pablum.

One of those, who had an injured wing, remained with us while the others went to the Belize Bird Rescue for rehabilitation and eventual release. Over a period of four years, 13 of our rescues went to Belize Bird Rescue, while we kept some who we felt would do better by remaining in our aviaries. We always lectured the thieving youth on the importance of leaving baby animals with their mothers, and with threats to report them to the Forestry Department, they eventually ceased coming.

Baby parrots lining up for "pablum"

However, the local kids were not the only nest robbers. We knew of a border official at the Belize-Guatemala border who supplemented her income by robbing nests. Through a person known to us who collected birds by hook or by crook, in 2007 she told us of two blue-headed Mealy Amazon chicks, sub-species that occur only in western Belize and Guatemala Petén province. The border official had taken them from a nest near the Guatemala border to sell to this person, who did not want them. We said we would take them even though we had to pay quite a hefty price for them.

The same official had previously sold this person three robbed toucan chicks, which all eventually faded in her aviary. We were shocked to see how young these blue-heads were, with parts of their bodies still without feathers.

Blue-head chicks

The only way they would take food was for Hugo to put the pablum into their crops with a special bird feeding catheter. They thrived under our care and grew into beautiful adult birds in the safe environment we provided.

Blue-heads all grown up

It was fairly easy to tell which parrots were compatible and thus decide who lived with whom. One day a smaller White Fronted parrot landed on the side of the aviary, obviously an escapee from another captivity. He definitely wanted to return to what was familiar to him, so without any trouble, Hugo was able to pick him up and house him with a female of the same kind. That incident further added to our belief that once compromised by human captivity, one needs to think wisely about which birds can successfully be released back into the wild. It is easy to become an eco-imperialist in a country were so many people have little regard for their wild heritage.

***

An incident we found very distressing was the inexcusable neglect leading to starvation and necessary euthanasia of a young howler monkey that had been in the possession of a socially prominent family. They had received this young monkey from the same nest robber who stole the Mealy Amazons. I had occasion to frequently visit this home, and watched the monkey held captive by a chain, change in two years from a playful little creature to an unrecognisable distorted apparition that could hardly move and just lay in a collapsed heap of fur on the ground.

As I noticed his decline, I took the liberty of frequently suggesting to these prominent people that their monkey was fading and suggested ways of improving his life: including the need to remove the collar that was restricting its large larynx used for "howling". I was always politely thanked and ignored.

Having been back in Canada for a while, I was shocked on my return to see the monkey's deteriorated condition and suggested to the lady of the house that the monkey was sick and something needed to be done. She told me that a local vet had diagnosed it with arthritis and that it was going to the zoo, but she could not tell me when. As I knew that a two-year-old monkey in a hot climate to have arthritis was next to impossible, I persuaded her to allow Nikki to fetch him. Within ten minutes Nikki arrived and I took photographs as recorded evidence of what Hugo said was the worst case of cachexia he had ever seen, and most certainly would have resulted in charges of animal cruelty had it happened in a more civil society.

A newly arrived wildlife veterinarian from Europe took X-rays that showed every long bone in the monkey's body had spontaneously fractured due to demineralization from malnutrition. Mercifully it was euthanized. I tactfully informed the said lady that the problem was not arthritis but malnutrition and suggested she not get a monkey again. She agreed.

A year or so later, she could not resist telling me that they had acquired another baby monkey, so infant that it was still kept in their house. Quite alarmed, I felt it was necessary to give her the autopsy report of the previous monkey, written by Isabelle, the wildlife vet, with the hope that it would convince them to surrender this baby monkey to "Wildtracks", a primate rehabilitation centre in northern Belize, but to no avail. I then managed to take a photograph of the infant and forwarded it to Isabelle. After that, the appropriate authorities set a confiscation in motion, which happened quite quickly in light of the cruelty inflicted on the previous one. That this action was possible at all happened because the Minister of Forestry at that time was a conservationist with integrity.

Chapter 14

If you will have men who will exclude any of God's creatures from the shelter of compassion and pity, you will have men who will deal likewise with their fellow men.

Francis of Assisi (1181-1226)

It was a few weeks before I left on a trip to England in 2008 to visit my niece Liesl and meet up with my sister Idille from South Africa that a few local teenagers brought a young coati to our gate. I had never seen one before, but she was the cutest little creature and I knew without a doubt that I could not leave her to a certain fate.

Mimi, our first rescued coati

She was old enough to eat on her own so we installed her in one of our earlier birdcages, with the appropriate internal configurations for her comfort. I did not intend to keep her but the zoo declined her as they had enough of their own and an American woman who lived in the bush to rehabilitate monkeys, reneged on her agreement to take her. As Mimi could not be abandoned on her own in the wild, there was no option but keep her.

Coatis, or *quash* as they are called in Belize, are members of the raccoon family. However, they remind me more of lemurs, with their longer hind legs and their tails carried erect. They are diurnal and their long claws enable them to be terrestrial and arboreal. They eat almost anything in the line of eggs, bugs, fruit and even small birds. It did not take long to fall in love with Mimi. Coatis grow tame and affectionate very quickly. I always marvel that animals can so readily form a bond with humans, more so than they do with other animal species. In no time Hugo set about building a walk-in wire cage on the stoop, large enough to accommodate branches, hanging toys and a sleeping box.

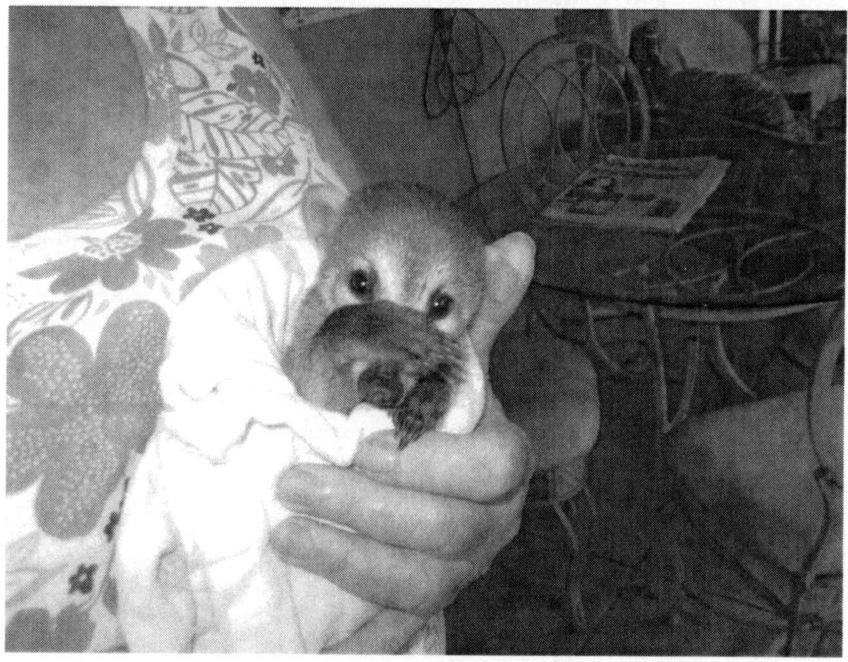

Another baby coati

Since it seemed she would be with us indefinitely, we had her spayed to prevent the frustrations and aggression that generally comes with sexual maturity; the time when locals who have them as pets, abandon them in the bush where they have to fend for themselves without having been taught how to. Her regular daily diet consisted of cat chow, fresh fruit, cooked or raw eggs (that she ate shell and all) and the odd captured cockroach, or any bug that ventured into her domain. As her cage was raised by a few inches off the stoop, a few times little ground doves would venture in with only the tell tale evidence of a few remaining feathers.

Two years after Mimi's arrival, another coati was brought to our gate, a real baby who could not yet feed on her own.

There was no thought of turning her away. With a doll-size feeding bottle and infant formula, it did not take long to nurse this little creature back to good health. She was named Angel. Not long after, as I was shopping in a local Chinese supermarket, I heard the by-now-familiar high-pitched chirp of a young coati.

I followed the sound down the isles until I found the source sitting on the shoulder of a Hispanic teenager shopping with her mother. Of course I had to engage them in conversation, and was told this one had bonded with the daughter and slept in her bed. I subtly suggested that they surrender the animal to me and assured them I had the facilities. I stressed the difficulty of keeping a coati, especially when they reach maturity. The woman said it was already a problem as the school holidays would be over soon and she would be left with the responsibility. A week after giving them my phone number, they called and Mona was added to our menagerie. She was quite a precocious little creature and loved human attention.

Leo came next, also very young. By now Hugo had built a second walk in cage on our stoop. Mimi hissed at the babies and was not willing to share her territory. I was able to "potty train" Angel similar to the way one would with kittens, but with more coatis and their individual smells arriving, the potty system eventually failed. We hooked up a garden hose and the floor was flushed into the garden many times a day. Needless to say, the tropical plants flourished at that end of the stoop.

A year after we got Angel, Mona and Leo, two more babies came our way. One was hauled out of a guy's pocket at our gate and was extremely dehydrated and underfed. We told the fellow to come back in a week's time and if the coati was still alive, we would give him fifty dollars.

Coatis growing up

Of course he made sure to come back, and fortunately the baby did survive. With a lecture that what they did was cruel and illegal we threatened to report them if they came by again: after that they stayed away.

Alan, an Englishman of independent means, and his Canadian wife Dolly, who had been globetrotting wildlife photographers, had recently settled in Belize. They bought a vast tract of bush in a remote area (a previous farm gone to seed), which they brought back to life and turned into a wildlife sanctuary they named "Witzoo". They went to elaborate security measures, which were necessary to keep out poachers and other undesirables.

We were thankful that they had agreed to take all our coatis where they would have the best of both worlds; freedom in a safe environment and still benefit from human contact if they wanted to. It was therefore with relief and sadness that we parted with them when Witzoo was ready to accommodate them.

Freedom at Witzoo with Dolly

Upon maturity, the females found mates in the bush but kept up their daily visits to the house with their litters tagging behind. The Grahams continued to accept distressed coatis.

***

At the end of 2008 we adopted two more Westies. My nephew Robert was moving to Australia from Denver and we agreed to take his dogs as they would not have withstood the stress of a long flight and quarantine without water. In December Robert flew to Belize with Maxi and Millie with all the required health certificates. We were in the Belize Agriculture Health Authority office at the airport and able to see Robert in the immigration area. The health certificates were handed in at the office. After perusing them for a while, the official told Hugo that he would have to fine him one hundred dollars because the certificates were not filled in correctly.

Hugo ignored him, as by now we knew he was angling after a "blue-boy". Robert had a small plastic box with a few rubber dog toys, which were duly confiscated. No doubt that particular official was also angling after a"blue-boy". When Robert eventually emerged, he told us that he had to pay USD$250 duty on the dogs. That was highly irregular but what do unaware gringos know about all the scams.

Chapter 15

Do not lay up for yourselves treasures on earth, where moth and rust* (and mildew) *destroy ...

Matthew 6:19

From the peace and quiet of our stoop we enjoyed watching our garden grow and the various little creatures that took up residence with us. On the stoop itself, tree frogs lived in windowsill corners or in pot plants. One took up residence in the tubes of a large wind-bell; the clanging pipes alerted us to its exit after dark and its re-entry in the early hours of the morning. One morning, a frog literally screaming from inside the wind-bell caught my attention. I could see the tail end of a small snake dangling out of the top of a pipe. A whack with a broom dislodged the snake, which let go of the frog, whose injuries did not seem fatal but it understandably moved away.

Fat-tailed geckos multiplied above the wooden ceiling boards and crawled out onto the walls after dark to feed on mosquitoes and other nocturnal insects. It was not long before the geckos moved into the house and we found their little eggs in unexpected nooks and pockets, including in curtain folds and my sewing boxes. We did not mind their presence as they performed an important task, but their little nocturnal droppings that landed everywhere, were a nuisance. However, we felt their droppings did not outweigh their benefits. They seemed quite territorial as we could recognise certain ones on the same walls every evening, scurrying behind picture frames when they sensed danger. The hatchlings were defenceless and we found many casualties around the house. The worst was when I saw a fat adult on the bedroom wall chase a young one and gobble it up. We were quite dismayed to see them as cannibalistic.

***

I was always on the lookout for cockroaches and every now and then I would find one in a drawer. Their furtive movements, even from small hatchlings, seemed to be evidence that they knew instinctively that they are universally hated. I got into the habit of rinsing all the plates and dishes before we used them in case a cockroach had left invisible footprints. One day I opened a small cabinet in which I kept crystal glasses (always placed upside down) and I noticed a cockroach: I immediately grabbed the Raid. When the insecticide fog cleared the 'roach was nowhere to be seen until I started unpacking the glasses one by one and much to my astonishment, the 'roach was hiding underneath a glass. The fact that it could lift the glass was astounding.

For quite some time I had noticed the frequent appearance of little bugs around my cats' feeding tray, which was placed on top of an old mahogany tallboy in the bedroom. Even though I kept dispatching them with a whiff of Raid, they kept reappearing. One day when I was putting away some of Hugo's clothes, I noticed a few dead 'roach carcasses in a drawer. Then I found empty egg cases, alien looking oval black pods that can contain up to thirty baby 'roaches, miniatures of their parents and equally obnoxious. Frantically I checked through every drawer and found evidence of 'roach breeding and death everywhere. With a feeling of revulsion, I tossed all the clothes into the washing machine, turned the tallboy upside down and emptied a can of Raid into every wooden crevice.

***

During the first rainy season in our new house, when the humidity was at its height, I noticed green fuzz on the armrests and drawer knobs of the wooden furniture. Lilly calmed down my consternation by telling me that it was only mildew growing on body sweat, and that it would all disappear again once the dry season rolled around. After that, the weekly dusting was done with a dampened cloth and the problem was minimized. I also noticed fine white fuzz growing on all my wooden spoons and my rolling pin. We sawed the rolling pin into circles to string together for parrot toys and I replaced all the wooden spoons with plastic ones.

However, plastic had its own problems. Some Tupperware items delaminated and unless plastic was of a high quality, it dried out and cracked. White telephones and other white plastic parts of small appliances turned yellow in no time. Elastic perished and lost its stretch, which could cause embarrassment if one did not check one's underwear.

Mildew invaded everything. It grew on certain colours of my sewing thread, mostly the darker shades. Occasionally I found a jar of jam that had become mouldy even in the fridge. Powdered spices like mustard and pepper turned into hard unidentifiable objects with a covering of fuzz, and the flavour of herbs and spices dissipated in no time at all. I soon learned to buy fresh dried herbs and spices on trips to America or Canada, as one never knew how long they had languished on local store shelves. Once I returned home, I stored all herbs and spices in the freezer, where they lasted indefinitely in their original state. Mildew grew in the fine hair-like cracks of ceramic ware and on some Teflon pans. While visiting, Sandra asked me if my cast iron cookware was rusting; she laughed when I said they grew mildew instead. I was surprised to find mildew even growing on some of my lipsticks.

In the wet season mildew also grows on concrete walls, pathways, outdoor plastic furniture and even stoop cushions that have perspiration on them; therefore, maintenance is high and on going. Our house was painted with a high quality mildew resistant paint, but we had to power wash the concrete walkways around the house on a regular basis. This slowly eroded the smooth finish on the surface, erasing William's leaf imprints.

Most public and government buildings are not painted and have a permanent dirty look where mildew has taken hold on the rough concrete. The Belmopan Museum stood out like a sore thumb. The lime wash on the walls was constantly washing off, leaving ugly dark streaks that did nothing to stimulate any curiosity about what was on display inside. Instead of a good coat of paint that would last, new lime wash was periodically reapplied.

***

Rust was another enemy. Unless the stainless steel was of a high grade, it could not withstand the corrosiveness of humidity. My nephew was sure he could even taste the corrosiveness in the air. My stainless sewing needles, made in the United Kingdom and not China, rusted just at the spot where they came in contact with the black paper that held them in place. Not being a chemist, I could only surmise that it was a chemical reaction between the paper and the humidity. The spool part of my sewing bobbins rusted, staining the thread. My Swiss made sewing machine fell victim when a section of the shaft that holds the needle also rusted. I had to wrap duct tape around my sewing scissors where my thumb fitted through, as my sweat was causing it to corrode.

Our fridge (brought from Canada and had stood us in good stead for many years) developed rust on the upper and lower doors, to the point where I was sweeping flakes of rust off the floor every day. It also seemed to have lost its air-tightness, so we reluctantly conceded that it had to be replaced. A local Hispanic second hand consignment entrepreneur volunteered to take it off our hands.

We replaced it with a smaller capacity fridge from a local British-owned furniture store that was reputed to sell new appliances and not reconditioned "lemons" from America, as some other shops were known to do. The new fridge turned out to be not that ergonomically comfortable and we contemplated the impossibility of trading it in for a larger one while it was still under warranty.

Going past the consignment business, Hugo noticed that our old fridge was on display, looking almost new. The doors had been re-coated to look like the original. Feeling somewhat nostalgic, we bought it back and found that the recoating had also hardened the rubber door seal. We solved that problem by sticking weather stripping around the doors and we made sure to give it a hard push every time we closed it. Necessity is certainly the mother of invention and it was often a surprise to learn how resourceful some Belizeans were.

One such man was a Creole who had little education, but lots of entrepreneurial initiative. As the furniture store did not have a delivery system, this man parked himself outside as an independent deliveryman. The store manager was happy to send customers his way, so everyone benefitted. When we bought the fridge we too availed ourselves of his transport, which consisted of an ancient small lorry mostly held together with baling wire, and with a manual transmission. I had to drive with him to show him the way to our house while sitting on seats that had more springs than cloth showing. To prevent his lorry from falling apart over the potholes, it took almost half and hour to drive the two miles from the store to our house.

***

Having parted with all my fancy high-heeled shoes, I went to Belize with only sandals, including a few good quality pairs of Italian ones. One day when I reached for a pretty red pair, I noticed fine white lines on the black synthetic wedge heels. On closer inspection, I saw that they were cracks in the outside material and the white inside stuffing was peeking through. On another occasion I was walking in cork wedges, and felt one shoe give way. The wedge had simply cracked and broken in half.

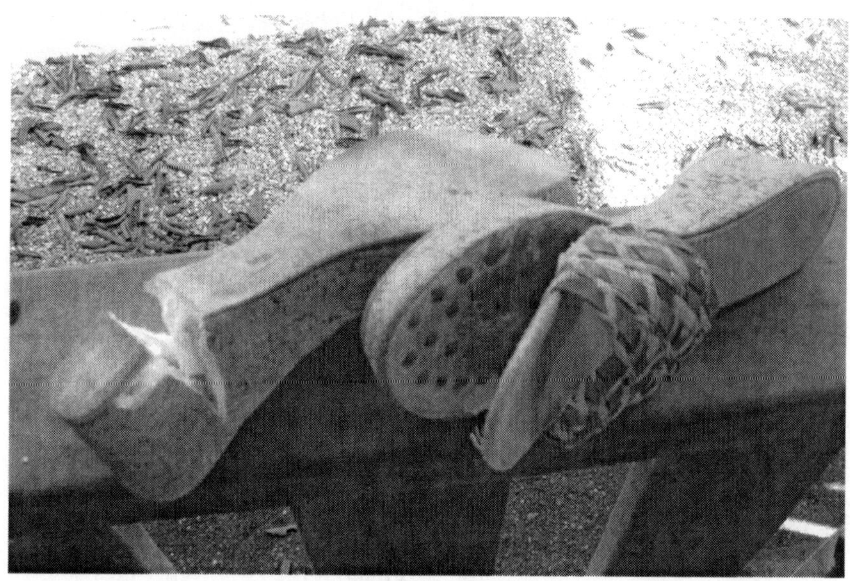

Broken shoe

One day Hugo put on his hiking boots and when he took his first step, the soles stayed behind on the floor. Birkenstocks seemed to be the only shoes that lasted. I returned from Canada armed with enough tubes of "ShoeGoo" that enabled us to keep shoe body and soles together. It also worked as patching material in other places.

Glue on shoe soles dried up

Early on, when buying envelopes, we discovered that the glue had already been moistened by the humidity and the envelopes were already sealed, and if not, the glue had dried up completely. The first time we posted a letter secured with scotch tape, the post office man ran after us, telling us we were not allowed to use tape to seal letters and had to use glue instead.

The humidity and heat caused all sponge rubber to disintegrate and our sponge earphones collapsed into a powdery dust in no time. The sponge lining between the headliner and the roof of our Nissan disintegrated after a few years until the liner was almost hanging around our ears. In asking around, Hugo was given the name of a local church pastor who earned a living as an upholsterer. He had the Nissan for a day and the vehicle came back with a ceiling that looked as good as new.

We soon realised that taking all our photos out of their albums at the time of our garage sale in Canada, had been a mistake. Without the protection of a plastic album leaf, the photos became very sticky.

***

A few years ago students from Yale did research in Ecuador[10] and identified a fungal species called *pestalotiopsis* that destroys plastic. A 2001 BBC news article reported that scientists in Spain made the distressing discovery of a common fungus in Belize that eats compact discs[11]. The good news was that it only feeds in hot and humid climates. Belize certainly had all those corrosive elements. We subsequently kept our backup discs in the freezer. I packed up my generous collection of classical CDs, along with my good shoes and shipped them back to Canada.

The worst lay ahead. Over time my cats showed signs of respiratory problems, especially Guava. As there were no proper veterinary facilities or laboratories, it was difficult to make a diagnosis. All Hugo had was a medical bag with basic medications. We had a blood sample taken at the supposed best veterinary practice in Belize City, who was responsible for sending it to a laboratory. However, they allowed the blood to coagulate, rendering it useless, even though we had paid for the service. The X-ray taken at the practice was also useless. Hugo then had an X-ray taken at the human medical imaging practice in Belmopan, which showed signs of fungal pneumonia.

Lilly had a barely functioning donated microscope, through which Hugo was able to identify Cryptococcus and both cats were put on anti-fungal medication. There were weeks on end that I had to force feed Guava while Mango seemed to hold his own somewhat. It was a distressing time that did not endear me to conditions in the country. One night in August 2009, at the age of twelve, Mango went into distress. I watched helplessly as his amber eyes stared at me while he died. Mercifully it happened quickly.

I was determined not to go through that heartbreak again, so in November I flew back to Canada with Guava in the cabin with me. Hugo's previous associate tried her best to save the cat but after a month her lungs were too badly affected and she was gently put out of her suffering.

Chapter 16

The proper means of increasing the love we bear our native country is to reside some time in a foreign one.

William Shenstone (1714-1763)

Once we had moved into our new home, I decided I needed some domestic help. The haze on the louvered windows created by dust and humidity needed frequent cleaning. The tiled floors needed mopping and I found it too hot to do the ironing, although with tropical clothes, there was little of that to do. In asking around, I had been advised that certain people were inclined to be more reliable and less lazy than others.

Miss Paula, the little Maya lady who had been the first cook at the feeding program, was without a job and her husband was beyond being employable. He was a rotund little man who could be seen plodding around the town, reminding me of the comic strip Elmer Fudd, but without the gun. I thought life was tougher for the older Maya generation, as lack of education did not let them keep pace with progress, such as it was locally.

Paula was eager to houseclean for me. I fetched her and took her home once a week, from Las Flores where she lived in a little shack near her two sons. Most of the residents in that area were of 1980-1992 El

Salvador Civil War refugee origin. Driving through Las Flores, one had to dodge all the kids, chickens and dogs running freely in the dusty roads.

Paula was well meaning but slow. I had to teach her how to use a vacuum cleaner. Even with such a modern aid, she could not get all the cleaning done and I was the one who mopped the floors. When driving her to and from home, I never got used to seeing the neglected condition of so many dogs in Las Flores. I would always gasp at the sight and Paula would only give a little chuckle.

After almost a year in our employ, there was just such an incident one day, when nearing her home I saw a bitch trotting in the same direction. By the time we reached her house, the dog had arrived too. She could hear the astonishment in my voice when she acknowledged that the dog belonged to them. Her grandchildren, as usual, crowded around the vehicle.

Then I saw a colourless, stunted little dog chewing on what looked like a stone lying in the dust. I rushed to look and I saw an almost petrified old chop bone. I asked her if they wanted the dog and she said, *"No, an deh is nada wan"* (No, and there is another one). I said I would take them, so holding them by the scruff of the neck, the kids put both of them in the back of the Nissan and I took them home.

Their little bodies were cold and almost lifeless, so we put them in a box with an improvised hot water bottle and Hugo gave them subcutaneous electrolytes. He commented that he had never seen puppies so neglected. They clung to life for a few days and then died one after the other. It was our experience with some other severely neglected animals, that they cling to life, but when they are rescued the fight leaves them when they relax, by which time they are too weak to survive.

When Paula came to work the following week, we had a talk. I told her about the need to have compassion, but she told me that they were poor and she could only afford compassion for her sickly granddaughter. I tried to explain to her that compassion is free, something that stemmed from the heart, and that she knew we were there to help her if she needed it. She said she was too ashamed to ask. I could not reconcile her logic (or lack thereof) that it is too shameful to ask, but not so to starve unwanted pets to death.

Shortly after this incident Lilly arranged a free spay and neuter clinic in Paula's area, but she did not avail herself of the opportunity to have her dog spayed at no charge. I was pleased that the Christmas holiday season gave me a convenient opportunity to end our working association as I found it hard to look at her and still feel benevolent.

I had a small succession of "once a week" domestic helpers after Paula, all of whom I grew rather fond of. They were honest, hardworking and intelligent women. As is common for many, they had been abandoned by their husbands and were left to raise their children without financial support. Even though Noelia had the courage to attempt to get maintenance through the court system, she was never successful. According to her, the court always accepted his excuses for not paying. Mireli used to relate to me how unlucky in love she was, because her husband had "sweethearts" and she was left with the responsibilities of caring for her three daughters alone.

For married men in Belize, to have "sweethearts" is a common and acceptable practise: not so of course for the betrayed wife or partner. One successful Belizean businesswoman married to an American, with whom we were acquainted, shared with me that as a little girl her father would sometimes take her along when he went to visit his "sweethearts". Both Noelia and Mireli, who were Hispanic, left their children with relatives and went to the United States on work visas with the hope of making a better living. The trade off between leaving their children behind and earning more money was a difficult one. They returned home before long.

My most recent help was a young Creole woman. It was unusual that Sulma was still unattached at the age of 32. She told me she did not trust men. Her father had abandoned her mother when she was pregnant with their ninth child and with the help of a brother she managed to raise her children. Sulma still became teary when she related the abandonment to me. Now her sister had three children by an abusive partner whom she was too afraid to leave.

Women are the marginalised members of Belize society and there is little recourse for them. What legal recourse does exist, again like most laws in the country, are seldom enforced. As one long time expat commented, "If there was such a thing as reincarnation, one did not want to return to Belize as a horse, a dog or a woman."

***

Hugo took a trip to South Africa in 2007 when his mother celebrated her ninetieth birthday. I was left to look after the yard and garden while he was away. I had my written instructions protected in a plastic bag, clenched between my teeth when I had to figure out all the ins and outs of driving the ride on lawnmower and working the pressure washer. I realised what hard work gardening was and in spite of Hugo's resistance up till then, I hired a garden boy.

Solomon was a young Maya student who was trying desperately to afford an education by doing garden work when he was not in class: after a few years he gave up the struggle. After two tries, he was eventually accepted into the police academy. After the standard four-month training period he proudly graduated as a policeman with Hugo and his father in attendance. His father, the patriarch of the family, and we established a good relationship over the years and all his sons worked in our garden at some time or other.

When we built the house, I reluctantly acquiesced to Hugo's choice of ceiling fans. Having only three blades, they did not move enough air and by now they were also squeaking, with the noise growing louder and louder in my head as the night wore on. While Hugo was away, the high quality ceiling fans I had coveted originally went on sale at the building supply store, so I bought replacements.

I sought out the Taiwanese electrician who previously did some work in our house (and whom I suspected had deliberately stolen Nyoni from a nest) to install the new fans. When Hugo returned home we had a garden boy and good ceiling fans, all that served us well in the long run. Even with the new ceiling fans, the discomfort of the humid summer months seemed to increase with each passing year so I started nagging for an air conditioner to be installed in the bedroom. Dave, the expat who had the refrigeration business, installed a "Made in China" unit that grinded like an old sewing machine. We had previously discussed the quiet quality unit that I wanted, so I had to assert myself to have the desired American made unit installed. He blamed Hugo for his mistake.

I was getting tired of all the dysfunction that was surrounding me. Most Chinese goods not fit for the North American market land up in places like Belize, which gives one an idea of just how bad they could be.

***

It was necessary to develop a sense of humour to survive all the strange idiosyncrasies of the country. So much of it was a comedy of errors. The police, who were stationed in the small villages, did not have vehicles but had to take the chicken buses to attend to any police matter. Their minister had given them vehicles that they crashed in no time and consequently not replaced. Having therefore to wait for the chicken bus, their arrival at the scene of the crime could be delayed by as much as a day after the fact. Just such a situation occurred when Noel, who lived on the Hummingbird, was attacked with his own baseball bat by some young thugs who cut his porch screens to get into his house. Fortunately he recovered from his injuries.

If we heard police sirens in Belmopan on a Saturday morning, where they did have vehicles, we knew it was never for a criminal or other emergency matter, but rather to lead the way for the cyclists who were embarking on a race. One could see the police, four of them in a four-door Hilux lorry, basking in the attention the sirens were drawing, as we all came out to gawk at the dozens of cyclists following behind, decked out in their Tour de France-look-alike cycling gear. Cycling was a popular sport and the narrow highways were the only place they could ride. Even though it was good to see so many Belizeans partake in healthy sporting activities (even to host an annual international cycling race) it was best to stay off the road during that time.

Sometimes traffic was held up by a Mennonite-built wooden cottage on a flatbed on its way to a new location. They would take up the entire road and one would have to wait for an opportune moment to pass, necessitating one to go off onto the dirt shoulder, if there was one. In the wet season, passing could be an even more hazardous undertaking, as one could sink away into the mud. One day I was stuck behind one of these moving cottages, with the amusing picture of a worker sitting in a chair on its back porch!

Driving was always a hazard. Cars turned in any direction without signals, or went through stop signs. Bicycles rode in any direction day and night, and almost no one had reflectors or lights. Those with darker skin blended more so with the night and made them even more difficult to see. If a gringo happened to hit a cyclist, it would be the gringo's fault. Therefore, we avoided driving at night as much as possible.

There were many head on crashes because locals tended to drive in the middle of the road. Nevertheless, staying to the side of the road was also hazardous because the roads were not maintained and chunks of asphalt broke away at the edges. My blood pressure would rise considerably by a common practise when acquaintances, going in opposite directions over a speed bump, would stop to have a chat and not care about the inconvenience to other motorists.

Adding to road dangers were the numbers of drivers who had bought their licenses or didn't have any license at all. One of my house helpers had a brother who was a taxi driver without a license. When Imogene-the-cook and I went to the market on a Friday morning, where I had to manoeuvre and dodge other vehicles for a parking spot, she used to say to me, *"Ah ku si yu neva bai yu jraiva lais."* (I can see you never buy your driver license).

The Department of Transport occasionally put out a pamphlet to remind motorist of the following:

> *<u>DEPARTMENT OF TRANSPORT</u>*
> *City of Belmopan*
>
> ***PUBLIC SERVICE ANNOUNCEMENT***
>
> *The Department of Transport advises members of the public of the following aids to traffic safety:*
>
> *1) Put on lights when travelling on the roads at night and during early morning, late evenings and conditions of dust, mist, fog and rain.*
>
> *2) Don't drive on the wrong side of the road.*
>
> *3) Don't overtake going up a hill or any incline.*

4) Don't overtake when you can't see clearly ahead.

5) Try to stop driving in the middle of the road.

6) Make sure vehicle rear lights are working.

7) Heavy equipment, houses and other oversized items cannot be transported nor can big farm tractors and other farm equipment be driven on the roads after 6:00pm.

It was bad enough that the roads were in such poor condition; bloated contracts and lack of maintenance were the chief culprits. So it was bad news indeed when a group of bandits from Guatemala came all the way to Young Gal Road near Belmopan and hijacked one of two $350,000 government owned road graders. For good measure they blindfolded and robbed the unfortunate operator of his money and cell phone before tying him to a tree and taking off with the grader back to Guatemala. It was spotted from the air near a small village across the border, but it had disappeared for good by the time a ground search followed up. One wonders how the bandits knew exactly where to locate the grader.

The Mennonites are good road builders who have all the necessary equipment. The proof of their capability is evident in the runway extension of the international airport and in their own roads in Spanish Lookout. We could only speculate why the powers that be did not hire them to look after the highways, which certainly would have been to the country's advantage. A gringo, who had lived all his life in Belize, related to us that originally he was a road contractor, but went into another line of work as he refused to grease all the palms along the way, which would have left him without enough money to do a proper job.

There were frequent reports in the news of bloated contracts and scams. One dubious contract that received a lot of attention[12] was about a road in northern Belize. A payment was made to a contractor, a relative of a high-ranking politician, but the road was never paved. The media people tried to follow the trail but were unsuccessful due to obfuscation or avoidance, experienced by many when dealing with anything governmental.

***

The American made recycled school buses used for public transport were a further hazard as they drove hell-for-leather on the narrow highways. It was best to get past them when the opportunity presented itself, either when they slowed down at a speed bump or were flagged down by a passenger, which could be in the middle of anywhere. Whether they drove in the middle of the road or on the side depended on the condition of the road. Each bus company had their own colour scheme and one got to know which ones drove faster and seemingly more reckless than others. One regular green bus company on the Belize City route I referred to as the "green mamba", as it was nearly as dangerous. Once Hugo came upon a head-on collision between a bus and a dump truck on a straight stretch of road.

I never considered catching a chicken bus myself, but many younger and more intrepid travellers do, especially if they do not have their own transport. Michelle and George was just such a couple from Canada. They were in Belize for a few months to complete a project: where their bicycles could not take them, the chicken bus did. On one occasion they were waiting for a bus in Orange Walk (a district north of Belize City) and the bus was one and a half hours late, by which time there were about eighty people waiting for a forty-five-seater bus.

When it eventually arrived, it was already half full. George, who is taller then the average Belizean bus passenger, stood firmly in front of the doorway to ensure that he and Michelle could get on. He had not reckoned on the agility of the short Maya ladies who slipped under his arms or the people who clambered onto the bus from the back. Nevertheless, they managed to get on and just when they thought they were on their way, the bus stopped outside a house from whence a man emerged, and with the blessing of the driver, siphoned off some diesel. Some of the passengers objected loudly and George, raising his voice above theirs asked, "So, what are you going to do about it then?" which elicited more loud discussion about people being all talk and no action. Eventually the bus set off for Belize City, only to run out of fuel on the outskirts of the city.

***

We enjoyed a rather comical outing that took place on the Governor General's field in Belmopan in July 2012. We had been invited by the British High Commission to an open-air big screen viewing of the opening of the London Olympic Games, a celebration combined with Belmopan Day (when the city turned forty-two). The printed agenda included addresses to be given by the city's mayor and a government minister, as well as the playing of the national anthems of both countries, preceding the screening.

Upon arrival the invited guests of the High Commissioner were seated under a big tent to one side and the general folk under another tent on the other side. It was all civilized and reminiscent of colonial days as we were served wine, followed by a delicious buffet dinner prepared by the High Commissioner's able cook.

Usual for Belize, the event was running rather late. The government officials were adhering to Belize time and had not arrived when they were supposed to. Jackie, the Deputy High Commissioner who hosted the event, was very apologetic about the delay. When the mayor had finished his remarks, the minister had still not arrived. In all this delay, a local church choir not included in the program, made use of this agenda gap, and positioned themselves on the field, conductor and all, and rendered their own program of hymn singing.

The deputy High Commission couple, who were not at all religiously inclined, seemed quite baffled by this unanticipated addition to the agenda. When it was evident that the minister was a no-show, the choir was dismissed and viewing commenced. It was an enjoyable event that gave us welcome cause for amusement amongst the many daily irritations.

Chapter 17

Move to a new country and you quickly see that visiting a place as a tourist, and actually moving there for good, are two very different things.

Tahir Shah (b. 1966)

With the dry season came water restrictions. Without any advanced notice, the water would generally be turned off at the most inconvenient time of eight o'clock in the morning and come back on at about six in the late afternoon. During that time we would have to resort to our improvised emergency system of buckets. We had the foresight for just such an eventuality by positioning a small rainwater tank near the kitchen door and a few cleaned-out-five-gallon paint buckets. Eventually we tired of this system and installed a six-thousand-litre rainwater tank with an electric pump attached, which automatically switched on when the city water failed. Hugo had to ensure that the pump pressure was kept up by occasionally using a bicycle pump.

The dry season is also springtime, when the locals set the bush alight. In our first year in our house, our neighbour decided to burn his acre. Hugo had to stand guard with the hose to ensure that the fire did not come near our fence. That evening, a wind came up and blew the grass ash into the house. I was not pleased with myself for lacking the foresight to close all the windows as I tackled the sooty clean up throughout the house the next day. Every day was a learning curve.

Traditional Maya *milpa* burning seems to have been adopted by many locals who do not even have a *milpa,* but set a match to any patch of unburned grass they pass by. This springtime burning causes the air to be filled with grey smoke, at times hanging like a thick fog and obscuring the sun. During those times we had to endure the constant smell and eye and respiratory irritations, with charcoal bits landing everywhere.

It is sad to drive along the Hummingbird Highway after springtime to see tracts of jungle that had been set alight for no apparent reason; not to mention the fact that animal and bird habitat is destroyed in the process. We never heard a word of discouragement from the government, only to say that they were "uncontrolled agricultural fires".

During the spring of 2011, the fires were particularly bad. The large amount of dried up bush debris resulting from Hurricane Richard the previous year stoked the fires and caused devastation to vast areas of jungle and to many trees that would never recover. On NASA satellite pictures one can see how much of the Earth's bush burns. It is obvious that all that smoke (along with volcanic eruptions) fills the atmosphere with carbon, so one wonders why global warming proponents don't address that problem instead of seemingly go only where there is money to be made by using fear tactics on gullible people.

***

Lack of customer service was the norm in Belize. Customer parking was mostly at the back of a lot, if any at all, with spaces (closest to the entrances) reserved for staff parking, especially at the post office and government buildings. In dealing with certain banks and many other businesses, quite often queries could not be answered and the promise of a return phone call never materialised. We often had the impression that some customer servers where clueless. Customer phone services were more often than not taken off the hook when there were utility interruptions, leaving one to wonder how long one would have to wait for service restoration. When the air conditioning unit went on the blink on a Friday evening at the height of summer, one had to tough it out until Monday, even though the business advertised that they provided a 24 hours service.

A billionaire British Lord, with dual Belizean-British citizenship (who had an intricate international business empire), owned Belize Telecommunications Limited (BTL) an Internet telephone service company. He was a controversial figure in Belize as he allegedly gave the People's United Party PUP (one of the two local political parties) one million dollars, who, when in power, introduced laws that were financially to the Lord's advantage. This, of course, rankled many people, especially those who supported the United Democratic Party UDP, the other political party. BTL worked hard to maintain their monopoly and made it very difficult, through fair means or foul, for other competitors to start a business. BTL was reportedly the most expensive and one of the slowest Internet servers in Central America. Not only were the rates outrages, BTL blocked the use of Voice over Internet Protocol (VoIP) so people could not use Skype or any similar services. Even though Hugo connected to a Virtual Private Network (VPN), BTL was constantly one step behind to block one. To boot, very seldom did we receive the stated speeds and it took forever to download anything. Hugo frequently complained to the BTL office, which was more to vent his frustration than to expect any improvement.

In August 2009, the UDP government nationalised BTL. Most of us were quite pleased with this new turn of events, simply as the prime minister had promised that VoIP would be unblocked and we looked forward to being able to communicate with the outside world without having to pay their exorbitant rates. However, nothing changed. BTL remained one of the most lucrative businesses in Belize.

Then, in 2013, to everyone's surprise and delight, the prime minister announced that the state owned company had agreed to lift the restrictions on VoIP, admitting that the fast pace of communications technology necessitated BTL to liberalise. That did certainly not mean a radical improvement overall; Internet rates in Belize were still the highest at 30 percent of the average person's monthly income, whereas the rest of the Caribbean was at five percent[13].

***

We had to contend with frequent power failures. Not having lights was not the worst, as we had oil lamps, but having to do without fans, especially at night, raised my blood pressure and body temperature. There was many an evening that I sat on our stoop with nothing on but a wet bath towel wrapped around me. The frequent fluctuations and blips in the electrical current also caused damage to appliances and electronic equipment. A Chinese resident could repair televisions, and we had to have the tuner replaced numerous times and the motherboard once. The capacitor on the air conditioner had to be replaced several times.

What added insult to injury was the fact that a Canadian shareholding company owned Belize Electricity Limited (BEL).

BEL office

A Canadian friend, who had shares in the company, said he did very well from them. However, those living in Belize felt exploited, as the rates were high and the service was poor. Even the Belize government was dissatisfied with the state of electrical affairs and in a dramatic move expropriated the company. Though we believe in free enterprise, we were glad for this take over: after that we were all of the opinion that the electrical service improved noticeably.

***

Since not receiving acknowledgement of his letter of introduction to the Belize Veterinary Association in 2004, Hugo was surprised to hear from Dewey in 2007 that the head of BAHA wanted him to do a survey of all the practices in the country, with the intention of raising the general standards.

A local veterinary clinic

A few more years went by before Hugo received a phone call in November 2009 from the chief veterinarian at BAHA. The two of them were to inspect all the veterinary practices throughout the country. Upon asking Hugo what he expected as payment, he replied that all he wanted was his petrol and lodging costs covered. That seemed acceptable and they planned to get started in January 2010.

Another year went by before the man from BAHA contacted Hugo and announced that they were ready to start. However, the deal had changed; now they would use Hugo's vehicle and he would have to pay his own expenses and then submit a claim. It was not difficult to imagine how that arrangement would conclude. Hugo declined to participate any further.

We read recently that a small private university planned "the development of a school of veterinary medicine"[14]. With such dismal professional standards overall and a lack of quality education in general, we could only shake our heads. An agricultural training facility would be a far greater need in the country with so much potential in that field and so little knowledge.

Another pie in the sky idea that came to light was that of medical tourism, which was said to be consistent with "National Sustainable Tourism Master Plan for Belize 2030"[15]. It is a good thing that they are aiming for 2030, as they will need that much time, at least, to bring about anything of that nature, if at all.

We used the services of a Cuban doctor who practised in a small private medical clinic. We found his services good enough for non-emergencies, but he left much to be desired. Even though I had taken my emergency medication for an impending kidney stone episode one night, I still could have benefitted from a shot of morphine. After much difficulty Hugo reached the doctor to explain my predicament. His only comment to Hugo was, "They will pass."

I spent the rest of the night enduring what was equivalent to three difficult deliveries. I wished him the same experience and never went back to him. A non-medical Belizean acquaintance said he could provide me with morphine if ever I faced a similar predicament.

There were a few good specialists who were recommended to us by word of mouth, as referrals from general practitioners are not required. We were satisfied with the medical imaging practice, where I could make my own appointment for an ultra-sound (when I felt it necessary) and who accommodated Hugo when he needed a pet X-ray.

We were also pleased with a recommended dermatologist, Dr. Bradley, whom we saw fairly frequently, and whose scope was hampered only due to limited facilities. She referred Hugo to a surgeon to remove two small skin cancers. He did a botched job as both re-grew shortly thereafter, and Hugo felt he added insult to injury by charging him a full consultation fee for removing the sutures. After she relayed Hugo's complaint to the surgeon, he refunded some of the overcharged money. Dr. Bradley had to resort to a long and unpleasant, but successful topical treatment.

We found it convenient, considering all the circumstances, that one did not need a doctor's prescription for most medications. We had to manage our own health for the most part, which included an occasional call back to Canada for advice.

I fortunately never had the need to use a dentist in Belize.

A local dentist. Photo Sally Thackery

Hugo only went when absolutely necessary, and was reasonably satisfied with the one who was recommended by Paul. On one of her visits, Alice had a dental emergency and therefore had no choice but to visit the same dentist and pay the complete fee up front. Even so, he was unable to finish the necessary procedure and once back in Canada, her dentist had to rectify the situation.

The regional hospitals and the many rural health clinics all have shortages of funds, medical supplies, equipment and a host of other problems. Although medical treatment is free at the regional hospitals, in most cases the expertise is missing and one goes there at one's own risk. There is little choice for local people. With frequent reports of fraud in the management departments, it all seems like a big mess.

As it is, most wealthy Belizeans and politicians go to the United States for medical treatment. For that to change, the current medical systems must be replaced with better expertise overall and the local mindset will have to change to include honesty, integrity and service. As the saying goes, "Past behaviour is the best predictor of future behaviour."

The saying, "birds of a feather flock together" proved true for us in Belize. Our initial vision of blending into the local community turned out to be rather naive. The cultural divide is just too great. An expat is rarely accepted into the social, cultural or political life of Belize. Our friends were drawn from the diplomatic and expat community, some of the latter simply because we were thrown together through circumstances and not necessarily by choice. With ordinary Belizeans, many of whom we really liked, there remained a mutually respectful social distance. For some, it might have been partly due to the ghosts of the colonial past. Even though Belize is often described as a cultural melting pot, they have their own stratified social structure. The paler skinned Creoles, being amongst the social and political elite of the country, are regarded in some instances as "Royal Creoles". One just had to see all the skin-bleaching creams and hair-straightening pomades on the supermarket shelves to realise that many others aspired to be "royal". Like the world over, there are ethnic prejudices in Belize.

Chapter 18

No one is useless in this world who lightens the burdens of another.

Charles Dickens (1812-1870)

One day in September 2006, after nearly two and a half years in Belize, I received an email from Paul's American girlfriend Jennie, advising of a meeting at a private address in Belmopan, to consider starting an International Women's Group.

I had never been one to join women's organisations, but the overall scarcity of congenial social interaction in Belize spurred me on. I arrived at the address on Orchard Garden Street, not far from the British High Commission, where a large exuberant lady with a British accent welcomed me. I was ushered into the living room where about thirty other women of all sorts, foreign and Belizean, were seated. I was amazed to see I was not the only woman who had crawled out of the woodwork seeking connection with other like-minded beings. I wondered where they had all been hiding up till now.

As soon as Linda introduced herself and the reason for the meeting was explained, one could sense that she was a born organiser and motivator with a personality to equal her size. Her husband Paul worked for the British Foreign Office, and after a stint in Jamaica, had recently been posted to Belize. In Jamaica Linda had been part of an international women's organisation, and wanted to start one in Belize. All the women assembled were in total agreement, so there and then the Belmopan International Women's Group, thereafter known as the BIWG, was founded.

Linda was unanimously elected as first president and I accepted the nomination of vice; positions we both held for two years, after which Linda returned to the UK and I became president in the third year.

President Linda Lane and I as vice

The two of us became fast friends. The purpose of the group was two-fold; to create opportunities to foster friendships amongst women in that part of the country and to do some charity work to benefit local people, especially women and children who seem to be the most neglected members of Belize society: marginalised as one more fortunate Belizean woman put it.

We filled out forms indicating which activities we were interested in. Edith, a socially prominent Belizean and the only one fortunate enough to posses a swimming pool, generously offered it for the use of aqua fit, the one activity that took off right away and endured. We were happy for a regular opportunity to get relief from the heat and benefit from exercise at the same time.

Aqua-fun

Our first instructress was Jackie, an ex-navy diver from the United States. Two years previously, Jackie and her husband along with two small children moved to Belize sight unseen. They bought a jungle restaurant (on a friend's property) in a very remote but lovely part of Belize called Barton Creek Outpost, with an almost impassable road in the rainy season.

They learned about life in Belize the hard way. When she shared their hardship experiences with me, I asked her why they did not return to the States. She said they had been "Belized". That was a term one heard fairly frequently from people who had moved to Belize; only to face financial ruin for having believed all the glowing reports in tourist and expat magazines about "Living in Paradise" or coming to the country "sight unseen". Even if one was well informed, it was hard to avoid being ripped off here and there to a lesser or greater degree. I was happy for them to learn they had eventually been able to return to America eight years and two more children later.

Jackie made the arduous trek from Barton Creek to Belmopan twice a week, so it was understandable but disappointing when she announced one day that she would no longer be giving further classes. She was pregnant and the driving distance was becoming untenable. Not wanting to lose the use of the pool and disband our congenial group of women, I audaciously decided to take over the instructing. With memory and improvisation, we all survived the changeover and it became more like "aqua-fun". In 2009 Suzan, from America, took over where I left off.

Jovie, the Filipino wife of a German who worked on contract for the European Union (EU) as an advisor in rural development, instructed arm exercises by incorporating her knowledge of yoga. After seeing a little shrine in their house, I assumed they were practising Buddhists. Her husband's job had taken them to most parts of the world, and Belize was definitely not on their list of favourite places. They were more than happy to take up a post in Kazakhstan after their house had been broken into one night while they were asleep. When they left, Jovie told us (in no uncertain terms) that they did not want to receive any emails from any of us, obviously eager to shake the dust of Belize off their feet.

Not long after the BIWG became a reality, word came down that the Belize City International Women were not happy about our group and they were sending a posse of ladies to meet with our executive committee. We did not know any of them at that time and their mission seemed to some of us more like a hostile take over bid, as they wanted us to fall under their mandate.

Most of us did not see any purpose or benefit in doing so as our aims and objects were totally dissimilar, and we said as much. No doubt they were motivated by the monopoly mentality that is so pervasive on the all-encompassing political landscape of Belize. Some ill feeling lingered for a time in some quarters over our non-compliance.

The timing of the start of the BIWG could not have been better. The new American embassy in Belmopan had just been completed, including a safe walled in residential compound with air-conditioned housing, gardens, pool and amenities for diplomatic members, envied by many of us lesser folk. Since America is the top dog in the diplomatic world, all the lesser embassies followed suit in moving from Belize City to Belmopan. With that, social life in Belmopan blossomed, as did the membership in the BIWG. In no time we grew from 35 founding members to over 100. Some women came from as far away as Corozal in the north, Placencia in the south and San Ignacio and area to the west. We included not only foreign women, but welcomed any Belizean women who wanted to join.

Because Linda was attached to the British Foreign Office, the High Commissioner at that time saw her efforts as positive community building and hosted a cocktail party in our recognition; therefore, we had his blessing to use their clubhouse, The Pig and Parrot, to hold our monthly early afternoon meetings. After the reports were read and guest speakers completed their presentations, we would enjoy a social hour with potluck refreshments.

Our first order of the day, as a group, was to plan and organise fundraising activities. Our most ambitious and successful event was an annual themed dinner and dance held in the George Price Centre for Peace and Development, a modern air-conditioned community centre. Although the food was catered, we women provided ethnic desserts that literally added the international flavour to the event. The BIWG dinner dance perhaps became "the" social event of the year in Belmopan.

Another significant fundraiser was our annual Christmas Bazaar, which we set up at the local entrepreneurs' Friday market.

All of the items were donated and we limited the amount of second hand clothing as there were enough vendors providing those on a regular basis. Local people favoured household items and linens, winter jackets and any other objects that were not readily available in Belize. Our baked goods table sold out quickly, including the piles of sandwiches donated by some of the Belize women. Making iced layer cakes is not recommended in the tropics as the icing melts in minutes. For some unknown reason, the locals liked sandwiches, which were always the first to sell out.

Considering our cultural differences, which included being prompt or tardy, I felt our organisation ran quite smoothly and we all remained patient and civil with each other. We were bound by our common desire to be helpful in the community and we made enough money to support what we felt were worthy causes.

Christmas bazaar at the market

The wives of some of the ambassadors and senior diplomatic officials were a welcome asset as they became quite active in many of the BIWG charitable causes as well as their own; in particular was Dora from Brazil, Gwyneth and Kathy from the United States and Christina from Mexico. At the start, Dora initiated a feeding program for families living with HIV/Aids in the San Ignacio area. While the Brazilian ambassador and Dora were posted to Belize , she hosted our December meetings in the form of Christmas luncheons at their residence. As at all our meetings, we came with additional donations of non-perishable food items and toiletries to add, along with wrapped gifts, to Christmas care packages for these families.

The BIWG continued to provide monthly food parcels for the families afflicted with HIV/Aids, with distribution assistance from members who lived in that area. An annual Christmas party was hosted for these families at Hodes Restaurant, where the proprietors graciously made their open-air restaurant in San Ignacio available for the purpose.

Christmas party at Hodes

At last count there were nineteen families, which included 47 children, who received food parcels and gifts. The Red Cross and many other needy individuals received monthly donations of food. The BIWG was able to provide financial assistance to Gwyneth and Kathy who initiated a successful summer reading program for local children, a program that continued until the end of their postings in Belize.

Christina generously made her spacious residence available for our monthly meetings when The Pig and Parrot underwent extensive renovations. She also brought to our attention the Palm Centre in Belmopan, a home for thirty-seven mentally ill residents. As in most institutions in Belize, the Palm Centre relied on volunteers for staffing. On our first visit, the residents had just had their daily showers and obliviously walked up naked to the staff desk to be handed their very basic clean clothes for the day. After that we provided the necessary funds for Christina to buy underwear for the residents on one of her frequent trips across the Mexican border to Chetumal.

Shopping for underwear is quite limited in Belize. A Belizean woman told me that there wasn't any place where a woman could fit a bra before purchasing. With a diplomatic license plate, Christina did not have to brave the zealous Belize customs officials as the rest of us had to. The women continued to visit Palm Centre where they interacted with the residents and took them food and small gifts.

Another project that was initiated early on by the BIWG, was a monthly birthday party which included a cake and gifts for children whose birthdays fell in that particular month at Marla's House of Hope, a non-profit Children's Home. The home was staffed by volunteers and mainly functioned on donations. Their website describes the home as such:

> *We care for these children without regard for their race, religion, culture, national origin, financial status, or social status. Our long-term residential care is provided through a family setting centred around Christian love, example, and teaching. Where possible, we work with the Ministry of Human Development to reunite each child with his/her family...all of the children come to us from homes in crisis. These children are not orphans, but they are without a home because of abuse, neglect, and/or other difficult family situations.*[16]

The children enjoyed the attention, fun and games provided by the ladies who felt a special love for them. One of the sadder aspects was for those children who had to leave the shelter at 17 without a safe halfway house to go to. Especially the young girls, already vulnerable because of abuse and neglect, were easy prey for male predators who even included policemen. An American Christian couple we were acquainted with tried to establish a transition home for these girls, but were not able to overcome the bureaucratic red tape and had to abandon their vision.

One organisation in particular that came to our attention in 2009 and warranted on going support was a newly established safe house in San Ignacio, called Mary Open Doors (MOD), one of only two shelters for that purpose in Belize. MOD was founded by two Belizean women, Anna Silva and Marilyn Griggs, to provide shelter for women and children who were victims of domestic violence, an all too prevalent and largely unaddressed problem in the country.

The list of miscellaneous help provided by the BIWG over time was endless; ranging from thousands of dollars in medical assistance of all kinds for people who could not afford procedures; assistance for women who had to travel afar for cancer treatment; for dentures; educational projects and books; food and clothing; tuition fees for needy children; basic household needs for the elderly; donating money to worthwhile missions involving sick children; and financial contributions to various healthy activities for students, organised by other expat groups.

Joan, an expat from the UK, was the dedicated treasurer of the BIWG from the outset until her death early in 2012. She and I were like-minded in that we did not believe BIWG money should be spent on non-humanitarian needs, as that should be the responsibility of the Belize Government.

Belize receives so much aid and financing from a multitude of outside aid and non-governmental organisations that they seem to have come to expect others to foot their bills. There were many requests for such items; for example, picnic tables for some schoolyards and fencing for playgrounds. But the most astounding one was from one Belizean member who suggested that the BIWG pay to resurrect a long disused public swimming pool, donated and financed by some other entity in the first place, which the local powers that be had neglected to maintain, giving credence to the old adage, “Easy come, easy go”.

With time, as foreign service people moved on and less international women were available or willing to take on responsibility, more Belizean women stepped into executive positions. Their focus shifted somewhat to include financing certain non-humanitarian projects, but for the most part they have maintained the charitable support we originally initiated in 2006.

***

After the formation of the BIWG, Linda’s husband Paul started “Men’s Club”. They held their monthly meetings in the clubhouse the day after the women’s meeting. While we women worked hard, they played hard at activities like zip lining, boat trips, horse riding, river and cave canoeing and more. The Men's Club helped to create friendships amongst the husbands and men in general. Belmopan had changed from a quiet backwater city to a whirl of social activity and a place where newcomers of all sorts were welcomed.

We were certainly an eclectic group, with people from various walks of life and different parts of the world; from working in diplomatic service; on contract with the EU; or one of the many other foreign aid organisations working in Belize; to ordinary expats and Belizeans from the business community.

Hugo and I enjoyed many genteel dinners at the residences of the American ambassador and the British High Commissioner as well as their deputies. The Brazilian Independence Day celebration was always a grand affair with generous amounts of food and libation, with many of the Belizean political and social elite present. There were also private dress up parties, other any-excuse-to-have-a-party parties and sometimes "street parties" in the US compound. Generally known for excessive imbibing, some expats were not always on their best behaviour at some of these social events.

The Pig and Parrot clubhouse continued to be the once-a-week gathering place where people on their guest list could meet socially. Those who misbehaved were struck off the list. It was also the venue for many expats to gather for New Year's Eve potluck parties.

Since most of the embassies relocated to Belmopan in 2005, a few more Chinese supermarkets opened, one even called "The Mall". The selection of groceries improved immeasurably and we were finally able to get excellent fresh coffee beans from Premium Wines in Belize City. Our back stoop was a convivial place to entertain friends. On a few occasions we had a large group of people, but we preferred smaller intimate gatherings, starting on the stoop and ending around the dinner table. We mostly had to set the time for just after dusk, when the parrots' regular late afternoon squawking quietened down.

Friends around the dinner table

Friends for Christmas dinner on the stoop in 2008

Belize was like a revolving door with the various aid personnel and diplomats coming and going like passing ships in the night. Needless to say, during that time we met some interesting people and made a few good friends and acquaintances along the way. There were also some rather strange people, especially amongst the expats. It is not uncommon to find the flotsam and jetsam of the world in expat communities. Belize certainly had her fair share. We were all refugees from something or other: colder climates, dull skies, bad marriages, income tax, sometimes the law or just the world in general. It was said that many of "America's Most Wanted" were caught in Belize. We naturally did not meet any of the latter, as they would be the one's to keep the lowest profile.

During this time we became good friends with a young Israeli couple who had just arrived in Belmopan. Omer came to take over the management of a fish farm, the Israelis being at the world forefront of tilapia husbandry and therefore managers of choice. Included in their contract was a house, located between some shops just off the Ring Road. I was quite shocked that people from a civilized country would be given such a dilapidated little house to live in, where cockroaches dropped down from the ceiling onto their heads at night.

Yael, Omer's wife, was a real trooper and made the best of a bad situation. She found she was pregnant with their first baby just before their arrival. She had a bicycle for transport that she used to ride to attend the aqua fit sessions, in spite of her slowly expanding girth. Unlike many of the elite Belizean women who go to the US to give birth and then return home with an instant American citizen baby, Yael had no option but to have her baby in a Belize hospital. Another expat acquaintance, Simone, had more than one baby by choice in Belmopan Regional Hospital and she had no complaints at all.

Through Omer and Yael we met other adventurous Israeli couples. One notable couple managed a duty free diamond outlet in the tourist zone in Belize City. They fired a local employee after they were certain he was responsible for theft at their store. They were informed this employee arranged a "hit" on them. With Belize City having a reputation for violent crime and corruption, they did not wait around to find out, but made a hasty exit and returned to Israel.

Yael told me she survived the rigors of life in Belize only because she had a supportive and understanding husband who took her away from the "roach" house most weekends. In 2009 they returned to Israel with Yael and their small daughter departing ahead of Omer. Just before he was due to leave, Omer developed a painful foot infection of unknown origin that was not responding to treatment; severe enough to cause doubt at one stage whether he would be able fly back to Israel.

However, once back home and after extensive tests, it was discovered that through a skin puncture at the fish farm he had picked up a fish mycobacterium. Being a difficult and rare condition to treat, Omer would surely have lost his foot had he not left Belize at the time he did, as Belize does not have the medical expertise in treating anything but the mundane. We were happy for them at their timely departure, but we missed them.

Linda and Paul's term with the British foreign office also came to an end in 2009 and they returned to England. Tom and Monica, who were always generous contributors to the social scene at the US embassy, also departed. With their and some other social movers and shakers departures, life became less hectic.

Michel, on contract with the EU took over from Paul the responsibility of keeping the Hash[17] going, turning it into a more social event, which then included women. I chose not to be one of them. He also started the "Wednesday Wobble", a Wednesday-after-work-walk attempt at exercise whereby we walked around the Belmopan Ring Road, which was either dusty or muddy, depending on the weather. The "TGIF" was a Friday after-work-gathering at a different restaurant each week, but with not much to choose from in the line of restaurants in Belmopan, it died a fairly quick natural death.

***

A lot changed that year. My term as president of the BIWG came to an end and it was a relief that after three years, even though worthwhile, I was able to let go of that responsibility. I was feeling the effects of the local culture and other fatigue and decided to return to Canada for a while, taking my sick cat back with me. Our friends Hermann and Tracy generously allowed me to return to their tranquil cabin on the lake on

Vancouver Island. The setting was beautiful, peaceful and quiet and I relished being back in civilization and reconnecting with old friends. Alice lived close by so we enjoyed daily walks and other good times together.

During this time I re-established my Canadian residency, which meant I could not be out of the country for more than six months at a time. That suited me well as being able to anticipate a regular trip out of Belize made for happier times spent back on our lovely jungle patch. This pattern of going in and out of Belize would continue for over three more years.

In Canada I eventually had a home away from home with Alice. Her garage was lined with totes containing my more valued possessions and clothing appropriate for Canada, including that lovely bright yellow silk scarf I had bought at the local second hand clothing market for two dollars. During my absence, Hugo, who could previously hardly boil an egg, learned how to cook certain foods quite well. It was a case of either "do or die" for him.

As political guards changed internationally and people in the foreign service and the international community came and went, the camaraderie born out of the BIWG that had existed until 2009 gradually changed too. The Americans in general seemed less socially inclined than before, but who could blame them for not wanting to leave their comfortable compound. Their new ambassador and his deputy Jack, with whom Hugo developed a close friendship, maintained the same congeniality as their predecessors and we enjoyed many evenings in their company, as we did with the British High Commissioner and his wife Pauline. The BIWG itself lost some of its international flavour, which in turn created some disinterest amongst the international women.

***

However, there were always new friends to be made. Gabriel was an agricultural advisor from Uruguay who introduced us to *picanha*, a particular South American beef-cut best roasted over the barbecue. He also taught the main meat producer, Running W, how to cut *picanha*, so for the first time we had access to nice beef that became our entertaining staple.

In addition, in 2010, while I was on one of my sojourns in Canada, Hugo met Mark, an inveterate globetrotting oenophile from New

Zealand. He had recently arrived as an advisor on contract with the Commonwealth Secretariat. We were birds of a feather and soon became good friends.

Being in Belize without his wife Beverley, we liked to include him for meals over weekends when he and Hugo shared their taste for red wines and fussed over the progress of the *picanha* on the barbeque. We also enjoyed many good conversations over the dinner table. When Beverley visited in the summer, we drove down to Placencia to eat lobster during the week of lobster fest. By this time the previously dreaded road had been paved, making day trips possible, and Placencia had come into her own as a popular little resort area. The little Garifuna coastal village of Hopkins, 42 kilometres north of Placencia, had also gained popularity over the past few years with new tourist accommodations. However, the road there was still brutal and in desperate need of repair.

Jorge and Marta was a Spanish-Argentinean couple, Jorge being an engineer on contract with the EU, who joined us on many occasions. Marta could not speak English well, but with her exuberant personality it was not hard to understand her. Another friend, Arezoo, was the sweet Iranian wife of Hector, a Mexican diplomat.

Until Jack's retirement from the Foreign Office, his wife Linda, Arezoo and I had a standing weekly coffee date with Pauline on the veranda of the British High Commission residence where we caught up on the local scuttlebutt while being served coffee on Her Majesty's fine bone china. It was all very genteel!

Dolly and Alan, who had adopted our coatis, held a mango fest on their farm every summer where all and sundry would come from all directions for a potluck lunch and return home with mangos from their many trees.

***

One Sunday night in October 2010, while I was fortunately in Canada, Richard, a category one hurricane struck Belmopan head on with torrential rains and winds of between 90-100 miles per hour. In the greater scheme of things, those are not the worst winds, but when it blows for most of the night it is devastating enough, especially for the poorer people who live in stick-houses and wood-and-tin shacks. Many lost their homes; 30 percent of the citrus crop was lost and the government estimated the damage at US $18 million. Since our arrival in the country seven years prior, we had not experienced any hurricanes in Belmopan. Hugo had to weather this one on his own.

With a few days warning he secured anything that could become a missile, covered the double front entrance with a plywood storm door and then waited. The power was shut off early in the evening, so all he could do was sit with the dogs and an oil lamp for light while listening on the little wind up emergency radio to the cries of, "*Help, di waata di kohn een*" (help, the water is coming in) via cell phones to Love FM radio station from people who did not make it to hurricane shelters, until the station itself had to shut down. (Almost every one in Belize owns a cell phone). Hugo said the constant roar of the wind and missiles thumping on the roof were quite frightening. The wind blew for about three hours, after which there was silence for a short while as the eye of the storm passed over, only to continue again from the opposite direction.

Jorge, Mark and Hugo cleaning up after the hurricane

By morning the storm had passed and Hugo awoke to the stoop and yard covered in foliage. He had to hack his way with a machete through to the aviary. Miraculously the parrots were unscathed in their corner shelters; the house was intact; the roof undamaged; ten trees were uprooted and twelve were broken off halfway. The thick bougainvillea archway over the front gate was blown over, bending the gateposts in the process.

It took Hugo all morning to clear the fallen archway so that he could open the gate and get out of the yard. Our portable generator kept the fridge and freezer going until the power was restored a few days later. Mark was the first person to arrive to help Hugo clear the debris.

Nigel, Alan and Jorge also formed part of the work bee over the next week. By the time I returned home, all was back to normal.

According to the local press the hurricane had some comical moments too. Motorists had to wait while crocodiles and boa constrictors crossed the highway in some flooded areas. In addition, a hurricane shelter warden, who opened his shelter in a small village north of Belize City, locked himself in while in a state of inebriation. In the dark he could not locate the keys and was therefore the sole occupant for the night while the villagers had to flee to another shelter.

Chapter 19

Something is rotten in the State of Denmark

William Shakespeare (1564-1616) - Hamlet

My sympathy for the women of Belize stemmed from my knowledge of Mary Open Doors (MOD), the safe house for battered women and children and the difficult circumstances of my own domestic helpers. The high incidence of domestic abuse of women and children in Belize is largely ignored, and certainly never mentioned in the glossy magazines depicting the virtues of "living in paradise". As president of the BIWG at the time, I met Melissa Beske from the United States when she attended a BIWG meeting with Anna and Marilyn from MOD. Melissa, an anthropology doctoral student at that time, was in the country on one of her frequent trips to do research and fieldwork for her doctoral thesis on "Intimate Partner Violence in Western Belize."[18] She also became personally proactive by facilitating a support group in October 2008, called "Women at Work Support Group of Mary Open Doors" (WAW), which was also to help abuse survivors gain financial independence.

Apart from weekly meetings, WAW developed a craft division in which members used donated craft supplies to produce weavings and jewellery to sell at the local Saturday markets in order to raise money for themselves and to support the MOD shelter. Initial efforts of MOD/WAW were so successful that it was recognised by the Women's Department of the Ministry of Human Development as being the most outstanding women's group in the country during the annual Women's Month celebration in 2009.

According to Dr. Beske's research, 44 percent of women in Belize suffer from domestic violence; of which about 99 percent goes unchecked in a society that fosters restrictive traditional, cultural and gender-based expectations for women, often perpetuated by parents and grandparents. The political system is mostly run by men who are indifferent to women, resulting in limited protection and access to medical care, education and economic opportunities.

I thought this indifference to women was well illustrated in October 2013 in an opinion written by a columnist in the popular *Amandala* newspaper when he commented on the proposed amendments to Section 46 of the Criminal Code. He stated, amongst other things, that rape of a female by a male is less heinous than that of male on male, as the former is a natural act and of little consequence if the female behaved like a whore and that females most likely provoke the men. These comments drew much publicity and criticism from various quarters. The National Women's Commission responded:

> *We live in a society where sadly, several men see absolutely nothing wrong to abuse and violate women and girls. What is even worst is that these same men shamelessly speak about rape as being part of our culture.*[19]

The columnist responded that he didn't know what the readers found so distasteful. Readers' comments were also varied, some for and some against the notion. Women's groups working for change definitely have an uphill battle in the male dominated macho society of Belize.

Coupled with poverty, there are not too many healthy cultural or intellectual extra-curricular activities and distractions for teens in Belize, so courtships tend to start at a fairly young age. During her interviews, slightly more than half of Dr. Beske's informants stated that they had already moved in with their partners during their teenage years: some did so to get away from abusive childhood homes. As having children is culturally important to both males and females, having babies often starts during teenage years. Due to strong regional Roman Catholic influence and opposition to birth control coupled with male attitudes, it is not uncommon for women to have as many as ten to twenty children during their lifetime.

Many men in Belize are raised to be machismo. Culturally they are viewed as the breadwinners and females their subjects, and if males lose that power they tend to vent their feelings of inadequacy on their partners. Further noted is when women disobey their husbands; refuse to have sex on demand or ask them to wear a condom; question them about finances; neglect to have food prepared on time; or go somewhere without their consent are all reasons for abuse.

Victims of daily domestic violence develop a lack of self-worth, which, combined with the struggle for survival and nowhere else to turn to for help, see little option but to remain in their abusive situations.

A vicious cycle is perpetuated by the acceptance of male promiscuity, often justified because his woman no longer keeps herself attractive. It is all too common for older married men to have "sweethearts"; younger and prettier unmarried women who in turn see the older man as a potential provider.

The children from these affairs are likely to grow up in fatherless and abusive homes; the negative social consequences of which are evident in the gang and criminal elements of the country, with the murder rate being the third highest in Central America. In July 2010, Herbert Gayle PhD, an anthropologist from the University of West Indies, Jamaica, released a study titled, *Male Social Participation and Violence in Urban Belize*, which addressed this very issue[20].

Some of Dr. Beske's informants related their own sad experiences in this regard:

> *I ran away from home when I was 16, and that's when I met my baby's father. He was my cousin's friend. He said he'd hook me up* (take care of her financially and emotionally*), and I fell in love with him. He was a nice and loving gentleman at first. Brought me breakfast, lunch, dinner each day, and I stayed with his sister, and he would sleep with me each night. I thought we'd get married and have a*

good family and house. I got pregnant with his baby when I was 17. And then, he started to back off because he already had a wife and another woman, and he didn't want nothing to do with the baby, and then I wanted to run away from him because I was so angry.
Ciara 18: domestic violence survivor

My dad was angry when I married. He was very strict and lashed me as a child. When I met my husband, my dad was mad because now there was another man, and he couldn't do that any more. He thought my husband wasn't strict enough with me, so he started visiting our house and told my husband not to let me have so much control ... and then my husband started beating me. I felt it was my responsibility to carry on culture, and that I should respect and follow my parent in order to be a good woman, and that is why I put up with it. My mother told me, "yu don mek yu bed haad, yu gwine lie enna." (You made your bed hard, now you're going to lie in it - meaning: you take the consequences.) *That's what she had done, and that's what she expected me to do too.*
Melanie 32: domestic violence survivor

He (the husband) has one woman at home and one for fun. She (the wife) pressures him about it, he goes out and drinks, and then he comes home and beats her.
Dr. Pacheco: medical practitioner

An astounding aspect to the sweetheart phenomenon is the acceptance by many Belizeans of "Outside Families". A Belizean chap known to Hugo and Mark had lived in California for many years where he met relatives of whom he previously had only a vague knowledge. When he met them he was struck by the family likeness and therefore felt encouraged to return to Belize to connect with, not only his own direct siblings, but also his outside family; his father's children with other women. Later he said that no one felt the need to hide the relationships, but neither were they interested in forming close ties.

Another Belizean, with whom we were well acquainted, found out that she was related (because of an unmistakable family resemblance) to a man similar in age who worked as a guard at the British High Commission. Ann eventually started talking to him and discovered that, even though she came from the western part of the country and he came from the north, they had the same father. Together the two families numbered 25.

The most public example is that of a former prime minister who has an "outside family", who I was told, lives next door to his official family. It appeared that the "sweetheart" children did not have their father's last name, but did receive paternal attention and support and worked on his political campaign. I always wondered if and how the family dynamics affected the children psychologically. Moreover, that type of lifestyle, in my opinion, was not a good example to a small nation where similar domestic situations perpetuates so much abuse.

Even though domestic violence is a criminal act and a violation of human rights in Belize, women are reluctant to report abuse because of police, court and legal system bias. As with many other laws in Belize, there is a serious disconnect between the law and its implementation or enforcement, a problem also mentioned in Dr. Herbert Gayle's report. Some informants discussed with Dr. Beske their views on corruption and ineffectiveness of the judicial system:

> *Police do nothing because they don't care and don't think domestic abuse is serious. The police pass you on to Social Services, who do nothing. If the man's friendly with the magistrate in court, they're on his side. So people don't report; they just keep on going. When it gets rough, the man walks out, and she finds a new man, has more babies, and the same thing happens all over again.*
> ***Amandina 78: retired public health nurse***

> *The police usually know the man, so then they won't help, and she just gets beaten for reporting. So women don't trust the police and even advise others not to report because then everyone will know your business.*
> ***Reverend Santiago: school chaplain***

The lack of education and resulting poverty exacerbates domestic violence in Belize. Women most often have little or no money to help them get away from their abusive relationships and support themselves. The 2013 World Economic Forum[21] report on the Global Gender Gap ranked Belize 107 out of 136 countries, including that Belize was the lowest performing country from the region on female enrolment in primary education.

However, domestic violence knows no class or boundaries. I became aware of an almost unbelievable situation where fairly recently a young Belizean woman was granted asylum in the United States because her abuser was her father, a well-known and respected man in Belize. Another relative had just been elected to public office; therefore, making it highly unlikely that she would get any family or community support or have a place to hide, reducing her chances of survival if she was forced to return to Belize.

For the first 17 years of her life, the girl was under the complete control of her father who repeatedly subjected her to physical, sexual and emotional abuse: sometimes being joined by her mother. Her siblings made no attempt to intervene on her behalf and at times even encouraged the abuse. She was denied secondary education and her father controlled any earnings she received from prostitution he forced her into. She endured numerous rapes. She was denied any medical attention during the birth of two babies, both of whom he allegedly killed.

Rather than helping the girl to safety, her family and the relevant criminal justice system and other institutional officials repeatedly blamed her for her predicament. Thus, their actions normalized and enabled her worsening situation. The physical and psychological damage resulted in nightmares, little motivation to eat, and feelings of suicide.

The girl was unaware of the two domestic violence shelters in Belize. Even if she had, without any economic support she would not have been able reach them from Corozal in the north as the two shelters are Haven House in Belize City and Mary Open Doors in Cayo district in the west. She was trapped in her abusive situation by the material and ideological constraints forced upon her.

With the help of Christian missionaries, who gave her a safe haven and financial and emotional care, the girl was eventually able to get away from her abusive home. Even then, the father and his cohorts continued to stalk her and threaten her survival. It was only after the girl was able to leave Belize with the missionaries, under whose protective care she would remain in the United States, could she continue her education and live a life free from abuse.

It is extremely encouraging to know that MOD is continuing to expand. They've secured five acres of land and they plan to build an additional shelter to accommodate 20 more families. In addition, the former Women at Work, (WAW), support group that Melissa Beske started in 2008 has since diverged to become Women Empowering Each Other, (WEE), and they are continuing to grow and develop professionally.

***

If you say there is no such thing as morality in absolute terms, then child abuse is not evil; it just may not be your thing.

Rebecca Manley Pippert
International Speaker and Author

Domestic violence goes hand in hand with child abuse. Far too many children in Belize suffer from both sexual abuse and the Commercial Sexual Exploitation of Children (CSEC)[22].

I am familiar with a young woman who was sexually abused from the age of 5 to fourteen. Her first abuser was a female baby sitter, followed by a male who rented a room at her grandparent's house where she visited during summer. He would go into her room at night. Her older cousins, both male and female, molested her during the day. Male boyfriends of her relatives followed after the cousins. Even though she felt that what was happening to her was wrong, her perpetrators always told her that everything was good and that they loved her. They made her believe it was normal. Her story is the classic tale of sexual abuse of children in Belize. Like many other victims, she was constantly told that the past is the past and that she needed to move on and get over it. There was nowhere for her to turn to for help. She continues to struggle on her own to come to grips with a life she knows has damaged her psyche.

It was estimated in 2007 that 50 percent of complaints about child abuse, sexual abuse and neglect were withdrawn and not prosecuted.

Even though Belize has laws in place for dealing with sexual abuse of children, the judicial and legal sectors are lax in investigating and prosecuting cases. The 2012 Human Rights Report for Belize[23] noted that in many instances the authorities would not prosecute cases if the child or parent were not willing to press charges. One would think that any judicial system would prosecute any sexual crime against children that they are aware of: whether the victim or parents agreed or not.

The 2013 United States Trafficking in Persons Report[24] placed Belize second from the bottom, and further stated that child sex tourism is an emerging trend. Whereas one tends to think that abusers are generally foreign men, a 2007 summary released by the International Labour Office[25], revealed that in a survey done with the participation of 30 victims of this trade, only eight reported that their abusers where foreign men living in the country, whereas the majority were Belizean men. Only a third of the victims were living with their mothers while most had never lived with their fathers, and none of them had completed their schooling.

Many tourists off the cruise ships that call at Belize City are a source of clients for CSEC. The barrier reef along the Belize coast necessitates cruise ships to anchor a few miles off shore and passengers have to be ferried to shore in water taxis. It is common knowledge that sex tourists can sign up with certain tour guides who provide access to young girls and boys. The tourists who are active in this type of sordid activity know where to find this market on the Internet.

It is reasonable to assume that the children, in most cases, can't be accessed without parental knowledge. Some parents willingly prostitute their own children not only to tourists, but make them available for sexual favours to older local men in exchange for money, payment of school fees, clothes and other gifts. This is known as the "sugar daddy" phenomenon with the same sexist cultural acceptance as the practise of husbands having "sweethearts": topics not readily discussed in Belize as they are regarded as private matters. The sad truth is that poverty is a major cause of much of these practises, exacerbated by the number of absent fathers in Belizean homes.

One of the members of the BIWG, who picked up children from school to attend the summer reading programs, arrived at our aqua fit session one day quite disturbed. She wanted advice on what she should do after noticing a little girl, not older than ten or eleven, standing outside Las Flores Elementary School all pimped up. The young schoolboys were teasing her and when the BIWG member asked what was going on, the boys said the girl was waiting for a man to pick her up. At that time we did not know anything about "sugar daddies" and we suggested she tell the school principal. On later reflection I thought he probably knew about it anyway, since the practise was part of the Belize sexist culture.

Emphasising that children are not part of the Belize tourism package, the Special Envoy for Women and Children, headed by Kim Simplis-Barrow, wife of the present prime minister, along with certain other groups, are making efforts to bring public awareness and change to the problem of child abuse and sexual exploitation.

***

Gabi came to Belize in August 2008 to adopt a baby because her native country in Europe did not allow adoptions from East European countries and she had expat relatives living in Belmopan; therefore, Belize seemed like a good choice. She did not realise when she came that it would be a long stay and an emotional roller coaster ride.

Before Gabi left home she had telephone contact with a Mestizo woman from Bullet Tree in western Belize who wanted money in exchange for a baby. Naturally Gabi refused as it would have been highly illegal. However, after she arrived in Belize she was able to trace the woman, only to be told that the woman had given the child to certain Mennonites in exchange for windows.

News spread via the jungle telegraph that there was a gringo woman in the country in search of a baby. Not long afterward, Gabi received two different offers of baby girls and one of a baby boy from the southern area of Punta Gorda: those babies were born out of incest.

A few months later, through an intermediary, Gabi was promised a soon-to-be-born baby from a mother who was expecting her eighth child. Both the mother and her husband were drug addicts; therefore, it was reasonable to understand why the child would be given up for adoption. After the birth, Gabi waited with bated breath for the promised baby, only to be told six weeks later that the mother had changed her mind.

We first came in contact with the local mother shortly after we moved into our newly completed house. One weekend she came to our fence with a small child in tow. She wanted to borrow money to buy gas for her stove, so she said, with the promise to repay us on Monday. Hugo, being a nice guy, was glad to be of assistance even though I suggested the whole scenario seemed rather odd. Of course she never repaid us and when we saw her again, she was begging outside the Mennonite poultry shop down the road where she was almost a fixture with various unkempt little children in tow and one at her breast.

The poultry shop, being a busy place especially over holiday periods, was a favourite hang out for women hoping to scam gringos with pleadings about sick children, dying grandmothers, houses burned down or whatever they thought was a heart-rending story.

In this addict's case, the grandmothers were looking after some of the older children, but it was disturbing to see the little ones so unkempt and being exposed to a lifestyle of drug addiction and begging. I wondered out loud why the child protective services were not stepping in, only to be told that it was due to the fact that she belonged to a prominent family.

The following year, after many other empty leads, Gabi and a friend from the BIWG went to see an abandoned baby who was left with the grandmother. After two weeks of waiting for a decision, the mother agreed that Gabi could fetch the baby, only to change her mind a day later. The mother then disappeared and a disappointed Gabi returned to her home in Europe.

In August 2009, a year after first arriving in Belize, Gabi was back: this time for what seemed like a definite adoption. Christina had been instrumental in arranging a baby girl, one baby too many for a young couple in her employ. All that remained was for the legal adoption formalities to be completed. In the mean time, Gabi and her new baby daughter moved into an apartment to wait: and wait they did!

First, the social worker had to do an interview. It took her five weeks. Next, she had to file her report, which amounted to two pages and took her a further eight weeks to complete. Gabi felt she was on the receiving end of some prejudice, especially as she caused some resentment by benefiting in the adoption process from Christina's diplomatic influence. All in all it took eight months for social services to complete the adoption process. Finally, it was a very happy Gabi who returned to her home in July 2010, with her mission accomplished.

Chapter 20

One may smile and smile and be a villain.

William Shakespeare (1564-1616) - Hamlet

When we first arrived in Belize, we found Belize City to be a fairly benign city and the only danger was being hustled out of a few dollars. Although we were warned to stay away from Southside, we felt fairly safe walking in the downtown area and on South Front Street where our rental house was situated. Most of the crimes committed were by petty rival *yoot* (youth) gangs who were involved in turf wars. As they were poor and could only afford bicycles, shooting rival gang members with illegal firearms from bicycles was their modus operandi, and know as "ride-by" shootings. Another name for them was "bicycle bandits", and when a similar incident occurred somewhere else in the country, it was referred to as a "Belize City style" shooting. It was all such a tragicomedy.

The comical disappeared as the street gangs became more violent and crime escalated. By 2012 the murder rate in Belize was 15 percent higher than the previous year, with most murders occurring in Belize City. Even children became collateral damage in street wars.

An example of this viciousness was a 14-year-old boy who was clubbed to death with a two-by-four and left in a bus terminal. Even with my own gun for self-protection, I no longer felt safe driving to Belize City on my own; especially after a woman was robbed outside the same hairdresser salon I used to frequent. I could not be sure I would be quick enough on the draw in a similar event.

One morning Hugo and I went to Belize City for some unimportant business. Finding parking on the narrow congested streets was always a challenge. We usually parked in a small parking lot belonging to a bank on Regent Street, where the parking attendant (knowing he would receive a generous tip) always let us in. To get to a shop called Hofius, we had to walk past the law courts a block ahead, and then diagonally across a small park to get to the shop. Just as we neared Hofius, we heard rapid gunfire coming from the direction of the courts and saw people running in all directions to and from the courts.

My first instinct was to think which was the safest direction to run for cover if mayhem broke loose. Fortunately, there wasn't any further gunfire so we joined the curious who wanted to know what had happened. A rival gang member gunned down another gang member as he came out of court, in full view of armed police who were not quick enough on the draw and the culprit made his escape.

One morning, early in 2013, four gang members were found murdered in an uncharacteristically brutal fashion in a flat in Southside. Arezoo and I were on our way to the city for a girls' day out when her husband called her cell phone from the Mexican embassy, instructing us to turn around and return to Belmopan immediately. We hummed and hawed for a while before deciding to err on the side of caution. The murder had sent the city into a panic, resulting in wild speculation as to who would be the next targets. The whole city went into a Belize-style lockdown with schools and businesses closing for the rest of the day.

Major crime steadily increased throughout the whole country. According to the United Nations Office on Drugs and Crime[26], with an average of over 44 murders per 100,000 population in 2012, Belize was placed as having the third highest rate in Central America and sixth in the world. In what used to be quiet Belmopan, homicides doubled from 2011 to 2012. By 2011 the population of the Garden City had also increased from 7,000 when we first moved there, to 14,000. Apart from increase in the usual burglaries and thefts, home invasions, human smuggling, money fraud, sexual violence, drug activity, the ever-present corruption and other criminal activities became commonplace.

Being such a small country with an approximate population of 334,000, everything that happened in Belize seemed as if it was in one's own neighbourhood and in one's face. The sense of disquiet created by the high rate of crime was exacerbated by lack of confidence in the police. Many of them were criminals themselves, accused of corruption; beating; extortion; drug activities; armed robbery and even murder in Spanish Lookout and elsewhere. Unfortunately they undermined the few who were trying to be honest and above corruption.

It was hoped that the appointment of a temporary Canadian police commissioner would help reform the police department, as well as mentor a promising future commissioner. We heard from various sources, that due to opposition from the top brass who did not want a foreigner telling them how to run their department, the arrangement fell through.

***

Lack of confidence lay not only in the police, but also in the prosecutorial and criminal justice system. They received a failing grade from the American Bar Association[27]. The conviction rate for murder was shockingly low at only 10 percent. Local media frequently reported on cases where witnesses were too afraid to testify and too often known criminals walked free because the prosecution informed the court that the case files could not be found; therefore, the charges were dismissed.

Frequently the police or other witnesses developed "selective amnesia" so there wasn't any evidence to be given. Some of the reasons given for the ineffectiveness of the system were inadequate legal training for prosecutors in all departments. In some cases police prosecutors appeared in court without having received any legal training. Needless to say, corruption was also a contributing factor.

***

The pervasive culture of corruption, in our opinion, was responsible for what we viewed as the general chaos in the country. It was rife from the top down, involving hundreds of thousands of dollars, on occasion even millions, from the top political and social echelons, to a mere two figures from those at the bottom. Anyone wanting a contract or permit had to apply to the relevant government department, where the minister responsible would expect a "cut" for granting the permit, which in all reality amounted to a bribe.

There were constant reports in the media about fraudulent land deals involving ministers and their relatives; issuing of passports and visas by ministers to mostly Chinese, in exchange for thousands of dollars; corrupt customs officers; foreign aid money disappearing; government supplies of inferior quality purchased at inflated prices from relatives' businesses and thus bypassing other commercial enterprises. No one was ever brought to book. The scenario stayed the same, regardless of which political party was in power. The power of a few good men was not enough to change the system.

Revealing graffiti

***

Even though denied by most, monopolies were a big part of the corrupt culture of the country. Many lesser folk felt that the leading 25 or so families in Belize divided up the economic landscape between them. One family had the sole right to import asphalt and foreign wood. Another family had the right to transport certain commodities; trucks coming from neighbouring countries had to transfer their cargo at the border onto the monopoly trucks, even if their destination was only an hour or so further.

Electricity and telecommunications were private monopolies until the government nationalised them. Would be competitors were refused business permits. The best-known monopoly was that of the brewery and soft drink company. An acquaintance even had one bottle of Pepsi confiscated when crossing the border from Mexico, to protect the Bowen and Bowen Company who made Coca Cola.

Recently a young American couple on Ambergris Cay received all the necessary environmental clearance to start a microbrewery, only to be denied a permit by the Customs Department, the reason given was that granting such a permit, "Would open the floodgates on similar and other unique business ventures".

***

I heard students on a local television program discuss what they thought were the negative aspects of progress in their country. Their awareness was encouraging when they cited "resistance to change" as one of the main reasons. Maybe there is some hope for the future. A good example of resistance was hiring a highly qualified professional from outside Belize as the president of the University of Belize, a non-accredited institution, with the purpose of raising the standard and quality of education.

An audit required by the European Union to assess the millions of dollars the university had received in grant aid raised questions about the financial management; the Board of Management that was politically appointed, disregarded the rules and regulations. The ensuing changes for improvement that the president tried to implement were met with critical protests and political interference, which obstructed his efforts.

He resigned before his contract expired and the university no doubt will maintain their original status quo.

A friend, who resigned and left Belize in similar frustration and disillusionment because of political interference and incompetence, had returned to Belize after years of business education and experience in America, with the desire to help her country progress after the changing of the political guard in 2008.

She was appointed to a high position in a government organisation supposed to stimulate Belize's trade and investment. She established a trusted working relationship with the business community and did the groundwork for some positive changes, only to be thwarted by what she considered to be greed, ignorance and resistance to change in the political system. After leaving Belize, she shared her thoughts with me:

> *In my opinion, Belize will never change until it is forced to change. After working at ... I came to the conclusion that Belize is not an independent country. Belize may be independent on paper, but it is dependent in every other way. Maybe, if the Commonwealth, the EU, the NGOs and non-NGOs, and other countries stopped handing out charity, whoever is in power will listen to those of us who want to see Belize become a rich country. There is no reason why, in a country of 300,000 plus people, there are so many poor people. It is unconscionable ... There were enough studies done about Belize ... If they really wanted to help, they would send technical people to show Belizeans how to implement new ideas, help with planning and organising ... instead of providing fish for people, they need to teach the Belizeans how to fish."*

An excerpt from a Washington Times article on January 16, 2012 titled *Tale of Two Small Countries* by Richard W. Rahn[28], summed up the situation perfectly:

> *Cayman is rich, and Belize is poor. Why? Both are small Caribbean countries with the same climate and roughly the same mixed racial heritage, and both are English speaking British colonies.*

Belize should be richer: It has a larger population than Cayman (334,000 as contrasted with Cayman's 54,000). Belize has a much larger and more varied land area with many more natural resources, including gas and oil, and some rich agricultural land that Cayman lacks. Both have nice beaches, but Belize has the second-largest barrier reef in the world and also has Mayan ruins. Yet Cayman, with fewer points of interests, has done more to attract tourists. Back in the early 1970s, Cayman was as poor on a per capita basis as is Belize today. Both countries had ambitions to be tourist and financial centres. Cayman succeeded and has about six times the real per capita income of Belize. What did Cayman do right and Belize do wrong?

Most important is that Cayman had and maintained a competent and honest judicial system, which gave foreign investors confidence that their property would be protected. Cayman also has a very low crime rate. Unfortunately, the same cannot be said for Belize, where crime is often a problem. In addition, many judges in Belize are poorly trained, incompetent and, in some cases, corrupt. These issues cause foreign investors to consider higher-risk factors for projects in Belize as contrasted with Cayman.

It is obvious why Cayman is rich and Belize is poor, and it comes down to one word: governance. If Belize would clean up its courts, fully protect property rights and adopt the best economic practices of its competitors, it could quickly become rich. Those countries that are still relatively poor are poor because they have not put in place the necessary institutions, political structures and policies.

***

The excitement and euphoria of settling in a new country began to wear off; we could not help wondering why a little country, only a two-hour flight away from the mighty United States, was lacking in overall progress. Moreover, the stark contrast with Mexico just across the border added to the obvious. We began to strongly believe that apart from corruption, a contributing factor lay in all the handouts given to Belize by about forty foreign aid organisations in various guises, through which millions of dollars are funnelled into the country, creating an impression that Belize is a beggar-nation.

The European Union and the United States are perhaps the biggest culprits for creating this dependency on foreign aid.

Foreign aid

By way of illustration, the former gave Belize 42 million Euros over a period of five years. Amongst many projects, it was to diversify the economy, build roads and help big businesses and start up small businesses. It seems very little is done in the country unless an outside entity pays for it.

Starting in 2008 the United States began a huge handout initiative under the name of United States Central American Regional Security Initiative, referred to as CARSI. In 2012 they spent US$105 million on the Central American region, to improve security within the borders and counter the narcotics trade. Belize alone received millions of dollars in advanced equipment (much of it beyond their capability), training and other projects. Included in this largesse by the Americans, was the creation and training of a coast guard service.

In spite of the donated Boston Whalers and high-tech equipment, the shenanigans of the Belize Coast guard, as revealed in the media, had comic overtones. In 2009 a coast guard boat was stolen from a downtown pier in San Pedro on Ambergris Cay, when the crew left it unattended and went on the town. After an air and sea search the boat was found near Drowned Cays, but the engines were missing.

Also reported was that an engineer with the Coast Guard was charged with theft after stealing fifty gallons of coast guard fuel. On a night in 2011 a gang of thieves swam up to a coast guard boat, untied it from the pier and dragged it out to sea. Even though a guardsman fired at them, a getaway boat was able to tow the boat away. After another search, the boat was spotted two days later, thirty miles away, hidden in the mangroves inside a lagoon.

Most embarrassing of all was when, in 2013, the prime minister of Dominica, who was a guest of the Belize prime minister, was taken on an official boat ride to Hol Chan Marine Reserve near San Pedro, escorted by government officials and a police security detail. The coast guard was supposed to follow behind, but as they did not show up, the party went ahead without them. On their way back the party was intercepted and detained by the coast guard vessel. It was only after someone phoned the tourist minister's office, did the coast guard, without an apology, allow the VIP party to resume. The Dominica prime minister was understandably not amused.

There are other important providers of largesse, such as the Organization of American States; United Nations Development Program; Britain's Department for International Development; Canadian International Development Agency (now disbanded to become just Foreign Aid); Inter-American Institute of Co-Operation on Agriculture; Commonwealth Secretariat and many more. All, along with other nations like Taiwan, Japan, Mexico and Venezuela, have helped Belize to become dependent on others for handouts that amount to social welfare instead of commercial objectives.

Most studies left by departing foreign aid personnel are filed in an office drawer somewhere, some having lain there for twenty years, while the old status quo continues as before. Very little changed that we could see, crime continued to escalate and the people remained poor.

The ultimate in enabling Belize is, in my opinion, others picking up their trash for them. In what is known as "Walk for Belize", an annual event arranged by the Audubon Society, where embassies and other groups, including some civic-minded Belizeans, walk sections of the highways and pick up all the garbage littered along the road. So one can only ask, "Why would locals stop littering if they know others will be picking up after them?"

Sabina King, an expat traveller who experienced many local idiosyncrasies during her family's time in Belize, echoed our sentiments in her blog,

Oh, Belize. You are so beautiful on the outside and so ugly and decaying on the inside. Unless your government can provide its citizens with a positive example, you will be doomed to being rats grappling at each other for the leftovers. What a shame to a country full of rich diversity and beauty.

***

In 2012 the wife of a government minister and a socially prominent acquaintance, along with a policeman, called a meeting in our neighbourhood to organise a Neighbourhood Watch. A leader and a secretary were appointed; we filled in forms with our contact names and phone numbers to be distributed to everyone; and monthly meetings scheduled to be held; and that is where the Neighbourhood Watch began and ended.

One evening in January 2012, Hugo and I attended a party at the home of one of the few remaining socially inclined American couples, Donna and Barry, who chose to live outside the embassy compound, allowing them greater privacy. Their parties were colourful, noisy and generous; the latter included a surprise party for Hugo's seventieth birthday in 2010.

When we returned home four hours later, our three Shepherds did not come bounding up to the gate as they usually did. Instinctively we knew something was wrong. We found Daisy stone dead and in rigor mortis at the stoop steps, with evidence of her agonizing death throws in the gravel at the side of the house. Inja was lying unconscious at the laundry room door. Rosy was unscathed. By the rapid rigor mortis and other symptoms, Hugo knew immediately that they had been poisoned with strychnine.

Solomon, our gardener, found the evidentiary remains of tamales used to disguise the poison inside our front fence the following morning. Poisoning dogs was a known practise when criminals wanted to burglarise a house. The fact that Rosy was alive most likely prevented the attempt, but we were never really sure if burglary or sheer meanness was the motivation. Inja survived the night, simply due to his size, and made a slow recovery over the next days.

After that incident we did not allow the dogs into the front yard unattended. Henceforth we did not go out at night unless we hired a security guard to keep watch in the front; never being quite sure even then how trustworthy some of them were.

A matter of opinion

From our experience, Belize wasn't the charming place in the sun that we had experienced in our earlier years: we decided to put our property up for sale.

The culture of corruption and dysfunction; the subtle resentment towards foreigners; the increasing violent crime, targeting gringos and locals alike; and general lack of compassion for life, had slowly but surely sucked away my emotional energy and caused culture fatigue to set in with both Hugo and me. At first I felt very conflicted about giving up our own little tropical patch to return to our more orderly life of before. We had transformed an acre of jungle into a garden, the beauty and peace of which we enjoyed, as well as all the little creatures that shared it with us.

However, it was not an option for us to divide our year between two different countries. Hugo felt he would be able to ignore all the negative aspects, but gradually realised that he too was getting beyond the physical capability of the high property maintenance required in the tropical humidity. After the dogs were poisoned we both felt the constant vigilance removed much of the joy of living there. Belize was after all, not the benign country as portrayed in the many tourist promotions and international living publications. In my opinion it is in their own financial interests to portray the quaintness experienced as a tourist: not the realities of actually living in a country.

In general it is easy to buy, but hard to sell property in Belize (businesses even more so). Many people who would like to leave describe themselves as financial prisoners. Unless one gives realtors exclusive rights, which limits one's exposure in the market, realtors are not interested in promoting one's property. We were fortunate in that two different realtors, Hector and Ysenia, cooperated with each other in selling our house and shared the eight percent commission agreed upon, instead of the regular 10 percent most insist on.

Ysenia gave us the address of a justice of the peace who had to sign our documents. He was located in a dark little passageway in a building behind the market. We enquired at a regular looking office door if we were at the right location, only to be pointed to a butcher shop behind a grill in the wall. There was quite a line up of people waiting to be served at the window and we were not sure who was there to buy meat or who was there to get a signature. We were told we would have to wait until the butcher-cum-justice of the peace had finished cutting the side of pork.

After more than a thirty-minute wait, it was our turn at the grill. The butcher washed his hands, took our papers, and leaning on the big chest freezer, signed and stamped our forms. For once there was no charge for service rendered, and Hugo was so moved by this unexpected bonus that he gave the butcher a tip anyway.

Many of the friends we made were also due to leave: Jorge and Marta went to Spain; Mark was already back in New Zealand; Hector and Arezoo had just returned to Mexico; the High Commissioner and Pauline were due to retire soon; and Jackie and Nigel would be posted to Kenya in the new year. We were at last leaving too and we wished our remaining friends well.

***

Ode To The Newcomer

All the time we see them come,
Some are smart, some are dumb,
Some are black, some are white,
Chinese, Indian, Israelite,
Canadians, South Africans, often Brits
We tell them come and stay awhile,
Rent a house, Belizean style .
Belize is great, home's the pits.

Never rush, never jump
Or you could buy a rubbish dump,
"Oh no! Not me" they often cry,
I've been around, let them try,
I am too clever, worldly-wise
To buy a swamp or collapsed high rise,
I know the law, bring them on,
How often do we hear this song?

"I got money, want to spend,
Cash to burn, cash to lend"
"Slow down, SLOW DOWN" we tell them all,
But rarely do they heed our call
To hide their money, don't be flash,
Don't let them see your petty cash,
Take it easy, make no strife,
Come and try the easy life.

"I know, I know! I have a brain!"
Advice you give is all in vain,
"I'm not stupid, been around,
I've seen it all, there's no new ground,
Was in Belize, stayed three days,
Here's my plans, be amazed,
Bought some land, has a creek,
Gone home to sell, be back next week"

"Going to build, got so much space,
Belize will have to change its pace,
I'm getting old, don't have time,
My place will be a real gold mine,
I'll make a fortune wait and see,
My plans will work out faultlessly,
I'll start a business, make some money
A king in the land of milk and honey"

And so they come despite the warning,
Truck piled high, and spend all morning
At the border checkpoint, getting mad,
Thinking that they're being had,
"How can they charge? This stuff's not new!"
As they watch the customs turn the screw,
"My truck is dirty, used and old
I DIDN'T BRING THAT MUCH GOLD!"

Credit card reaches max
On environment and sales tax,

And their pockets full of hard earned booty
Have emptied fast on import duty,
So to the ATM they have to dash
When the Customs guy wipes out their cash,
It's just a setback, not too bad
They're in Belize so just be glad.

So jubilant they wend their way
To the jungle deep where big cats play,
Where mozzies bite and scorpions sting
And the bush's thorns large scratches bring,
To the forest damp where mildew grows
That rot your 'electrics', shoes and clothes,
But they're not daunted, they are strong,
How often have we heard this song?

And oh! What joy! They find their neighbour
Can build their house, and do hard labour,
Can chop their bush, can plant their trees,
The smartest man in all Belize,
There is nothing that he can't do
Given cash and tools and wood and glue,
A house by Christmas, won't take long,
How often have we heard this song?

But cash aplenty are his needs
To start construction, plant your seeds,
A new machete to chop your grass,
To build a road so you may pass,
He needs material, steel, and screws,
Cement and block, no time to lose,
"Gimme dalla, gimme kwik,
Bai mee hamma, saa an brik."

He has a cousin, wife and brother's son,
His auntie's uncle's sister's one,
His whole family will lend a hand
To build your house and till your land,

To cook your meals and wash your clothes,
To guard your house whilst you doze,
It just takes money, little bit
On little bit and bit and bit.

Soon those bills are getting large,
The money pit's not free of charge,
The credit card is getting worn,
Our new arrivals look forlorn,
All the while demands for cash
Are diminishing their money stash,
The materials that came were wrong,
The nails too short, the steel too long.
And wondrous neighbour soon forgot
He told you he could do the lot.

Excuses and evasions come,
The started work was never done,
"Son too hat, ih rayn tu haad,
Mi granny ded, dehn bon mi yaad,
Foot payn mee, Mi house jrap dong,
Kaa done brok, Laisn gaan,
Pikni sik, he very ill
Gi mi mony to bai pillz,
Polees kech mee, Ay don no rang."
How often do we hear this song?

And soon they meet officialdom,
To Belmopan they must come
To show their passport, license, form,
To stay in a land of sun and warm,
Your residency they want to thwart,
Your retirement plans may come to naught,
Immigration takes too long,
They've lost your file; your paper's gone,
Pay more cash to extend your stay,
Wait one month or two they say.

The mall's not built, the shops are bare
Of modern goods, except Chinese fare,
No bowling alley or cinema,
Was it wise to come this far?
The roads are bumpy and have big holes,
My pickup truck is looking old,
There's no Big Macs or KFC,
Italian restaurant ceased to be,
The power's hardly ever on,
How often have we heard this song?

The rains did come, the land did flood,
The building site has turned to mud,
The lush mangrove that I cut down
Has caused my coconuts to drown,
And even though I had a plan
My huge Condo they want to ban,
It's not my fault, Third World you see,
I am foreign, they pick on me.

My neighbour's gone, my cash he took,
His auntie's wife could never cook,
My tools, my blocks, my roofing tin
Have vanished into air so thin,
Cement got wet, the sand was dirt,
My funds are gone, I've lost my shirt,
The gas was bad, my truck has seized,
I won't accept I've been "Belized",
My health has failed, not feeling well,
My worldly goods I have to sell,
Going back home, where things are normal,
Where rules are rules and life is formal.

And all because he didn't listen
To those well versed in his position,
He burned his bridges, came too fast,
And we all knew he'd never last,

We told him loud, we told him blunt,
NEVER pay your cash up front,
NEVER think you know it all,
NEVER think our tale's too tall,
And all advice that he forsook
To never jump before you look
Has sent him packing, pockets empty,
Back again to lands of plenty.

Be even though we are so smug
We know Belize is like a drug,
That if you come and stay awhile
You'll be swept in Belizean style,
A pirate's land, with pirate's luck,
We need their cash, their loot, their buck.
Where many fail, just some succeed
To fill that urge, that inner need
To live a life, exotic, free to feel
Jungle trail and sun and sea

So who can blame them, those who come
To try their luck with what we've done,
But sure as dawn on misty mornings,
The ones that fail, ignored the warnings -
Things are done here differently
To the things back home you wished to flee,
So bide your time, be at ease,
Time means little in Belize,
Its not that we don't know its wrong
But we do get tired of this old song.

By Jerry Larder[29]
(An expat living in Belize)

Chapter 21

You never know what events are going to transpire to get you home.

Og Mandino (1923-1996)

We would miss our garden hewn from the jungle and our parrots saved from unscrupulous nest robbers, but we were satisfied that we had sold our property to the right couple; Jimmy and Sally from America seemed like kindred spirits and they would pick up where we were letting go.

After hastily packing up our furniture and a few other belongings that had survived the ravages of heat and humidity, Lilly and Dewey jumped in to organise a garage sale of the goods we decided to leave behind. Like everywhere else, garage sales in Belize were popular. It gave gringos and locals alike an opportunity to acquire items from departing foreign service personnel and culture fatigued expats, who leave with less than what they had arrived with.

Now, ten years later, we had decided to travel back to Canada the reverse way we came, through Mexico and the United States. As we were returning with four dogs, flying was not an option. Our trusty vehicle would get us back. In spite of a bumpy life, it had hardly a rattle and never broke down. Our little trailer was taken to Spanish Lookout, where it was repaired and given a new lease on life. Hugo made a new plywood lid with attached locks; the wood from the original lid had been salvaged for other purposes a long time ago. Instead of buying and tying on a new tarpaulin, Dewey suggested gluing on a piece of kitchen linoleum he had to spare, with enough of an overhang to prevent the rain getting in. The pattern was even colour-coordinated with our orange Nissan Xterra!

***

The thought of travelling through Mexico was rather daunting to say the least. There were frequent news and even some second hand reports of assaults on and kidnappings of both Mexicans and tourists by drug gangs and corrupt municipal and federal police who were working for the cartels; even news items of corpses hanging from bridges near Vera Cruz and in other places bodies placed in plastic chairs next to highways. Consulates issued dire warnings to their citizens about the dangers of travelling in Mexico; however, we felt we were up to the task and good to go.

With advice from Hispanic friends who had experienced travelling through Mexico, we planned to travel on safer toll roads which traverse the central part of Mexico; avoiding Vera Cruz and the now-dangerous gulf coast route we had taken on our trip to Belize ten years earlier. Our Mexican diplomatic friend Hector, suggested we keep enough pesos on hand as it was likely that we would have to pay *mordida* to police at the frequent check points, especially in the environs of Mexico City. We were also recommended to overnight in auto-motels that were safe, inexpensive and clean and should not mind our dogs. I had read that the auto-motels were sometimes referred to as "love motels", as patrons could pay for rooms by the hour.

As Sally and Jimmy took possession of the house at the end of June 2013, which fell on a Sunday, and as Hugo needed Monday to close out all our utility accounts, Jackie, the deputy high commissioner kindly invited us to spend our last two nights in the high commission guest house on Orchard Garden Street. We were very thankful for their generous invitation, as it gave us a safe, cool and comfortable place to stay with the dogs while we readied for our departure, instead of in a single room in the Hibiscus Hotel. On our last evening, Jackie and Nigel came by with a bottle of gin and we spent our last happy hour together in the tropics.

On July 2, 2013 we said goodbye to Belize on a rainy morning.

Goodbye Belize - photo Sally and Jimmy Thackery

Inja, weighing one hundred and ten pounds, fitted comfortably in the very back of the Nissan. The three Westies were on pillows packed evenly across the back seat. (Lewis, who had made the trip down to Belize with us was now aged and diabetic). Millie was prone to motion sickness so I was prepared with paper towels and a bottle of Dettol to disinfectant and mask any smells. The trailer was filled with our personal effects, some household linen, dog food, extra water and a cool box to preserve Lewis's insulin.

After driving through the last water filled potholes that splattered our newly washed vehicle with mud, we crossed the Rio Hondo River into Mexico. The border customs and immigration officials were friendly and helpful as we completed all the required vehicle insurance and other border formalities; quite a contrast to the surly Belizean counterparts we had come to expect in most cases. We could hardly believe we were on our way, driving on broader and well-paved roads. It was amazing to us what a difference one border could make. However, we were filled with some apprehension too, not knowing what awaited us at checkpoints. I had visions of being threatened, shot at (as had happened to an American couple) or having our dogs and vehicle stolen.

Even though she had been dosed with Atravet, an antiemetic-tranquiliser, we were not far into Mexico when Millie vomited. I felt relieved for having had the foresight to pack clean-up supplies. With checkpoints, tollgates and potty breaks for the dogs, we anticipated many more stops on this journey. The Mexican Army manned the first checkpoint we came to. They did not have the same bad reputation as the police, so we felt fairly relaxed.

They gave our trailer a quick inspection to make sure we were not drug mules, and told us not to bother locking up as a customs inspection lay a few yards ahead. With the latter inspection behind us, we were again on our way. Our third and last inspection on our first day of travel was by the federal police, recognisable by their black uniforms. We tensed up when we saw them ahead. They milled around the car for a while and then looked into the driver's side window. Hugo gave them a friendly greeting in his broken Spanish. Inja gave a sharp bark and the policeman commented, *"perro grande"* (big dog). I hoped the sight of him would deter them from any ulterior motive they might have harboured. After a brief inspection of the trailer, they waved us on our way.

***

By early evening the rain was still falling softly and we had not yet seen any indication of an auto-motel. With the anxiety of finding accommodation with our pets during our previous journey to Belize still fresh in our memories, we were beginning to worry about finding accommodation. While peering through the rain into the dark, in what seemed to be the middle of nowhere, we saw a sign with bright yellow neon words that read Motel Xul-Kah. With a sigh of relief we turned into the driveway. The registration office was at the entrance of the driveway going in, where one had to conduct business through a small tinted window. Angelo, the young man on duty, was very friendly and showed us to our room. That is when Hugo informed him that we had *perros,* and we felt great relief that he seemed not to mind. We appeared to be the hotel's only guests.

The "love-motel" on our first night in Mexico

The brightly painted turquoise motel had two long rows of rooms facing each other across a smoothly concreted driveway. Each room had an adjacent carport with a heavy rubber-like black curtain to draw across, instead of a door. Even though the car and trailer could not fit in completely, it did afford a little protection from the rain, as Dewey's linoleum was starting to show early signs of wear and tear. The concreted driveway kept the dogs' feet clean and there was a lot of grass outside for potty walks. With great relief we were safe and sound for the night.

Upon entering the room, we were immediately overwhelmed by a strange strong smell we assumed to be some type of disinfectant. After a cursory glance, we realised this motel did not cater to regular travellers like us, but strictly to those involved in illicit trysts. Evidently such behaviour was no longer frowned upon, even in a predominantly Roman Catholic country. Hugo and I looked at each other in bemusement, and then we laughed. It was no use crying.

The bed was a concrete block in the middle of the room with a flimsy mattress plunked on top and covered with a tatty old bedcover. Under the cover lay one grungy sheet, still with a few spots of blood and two dirty pillows. A low tiled concrete bench in place of furniture flanked the walls of the room; a higher shelf opposite the end of the bed held a television, below which was a mirror in line with the bed. The television had only one fuzzy channel showing a Japanese war film in Spanish.

Near the window stood a black vinyl-covered, contoured contraption that was obviously for more gymnastic lovers. For those inclined towards voyeurism, a large window from the bedroom looked into the shower cubicle. The adjacent toilet had no seat. Built into the outside corner of the room was a small hatch, through which a delivery of sorts — maybe some tamale — could be made while the occupants remained incognito. Before long Angelo knocked on the door to inform us that if we wanted a "special movie" we just had to ask at the office.

Once we had recovered from our shock, we removed the one and only sheet and put our own linen and pillows on the bed and made ourselves as comfortable as possible for the night. We could hear the odd car arrive and an hour or two later drive out again. After an uncomfortable night, we were up early the following morning. The smell of the room had permeated our bedding and our few unpacked clothes. As we drove away we looked back once more at the motel sign. Next to the motel name was the give-a-way figure of a girl in a yellow bikini. Her body was black, so in the dark we had only seen yellow neon. I felt somewhat soiled, even a little traumatised after the fact.

It was still raining lightly when we started the second day of our journey. We realised we had not been too far from the town of Escarcega the previous night. Even so, we did not notice any other motels in the vicinity that could have been an alternative to the Xul-Kah. That made us feels somewhat vindicated for our choice of the night before.

Beyond Escarcega we were stopped at two different checkpoints. We were relieved that neither one required us to unlock the trailer for inspection. Even though Millie had been dosed again with Atravet, and would be every morning for the rest of the trip, she vomited twice this time.

After their potty stops, the dogs were always eager to get back into the vehicle that had obviously become their security, as everything else that had been familiar in their lives had disappeared. By midday we reached Villahermosa. It was sunny and hot, but we still needed to find a tarpaulin, as Dewey's linoleum was definitely starting to look tattered. Through all the congestion, we spied a Wal-Mart sign. They did not have tarpaulins, but I stocked the cool box with fruit and cheese and baked goodies for Hugo, who by now was hungry for more than gluten free fare that I required.

In trying to extricate us from the maze of downtown Villahermosa, Hugo decided the GPS was incorrect and turned where he thought the right route was. However, the GPS was right and he was wrong, so we drove in many circles until we eventually found our way to the highway. All the while I sat in silence biting my tongue and trying not to go crazy. Secondary roads and city streets in Mexico have many *topes* (speed bumps), some industrial size. When Hugo did not slow down sufficiently, the trailer hitch became detached from the ball on the Nissan's tow bar; it did not help that the hitch was slightly worn. The chain that held the two together prevented the trailer from careening into other traffic. Hearing the scraping noise of metal on tarmac was the signal to stop and reattach the trailer while the traffic was held up. I steeled myself against embarrassment.

***

By now our well-planned route had fallen by the wayside: because of Hugo's lead foot on well-maintained toll roads, known as *autopistas* in Mexico, we were making better time than anticipated. All we could do now was to keep going and hope to reach an auto-motel before nightfall. After three tollbooths, by late afternoon we reached the coastal city of Coatzacoalcos situated on the southern Gulf of Mexico. We had to travel quite a distance through the large government owned Pemex petrochemical sector before reaching the city itself and kept our eyes open for a motel sign. While travelling on a fly-over bridge, we spied the Privilege Motel below but we had to keep going until we found somewhere to turn around. We found ourselves in the city centre at the top of a hill, where we saw a Best Western Hotel. Not surprisingly they refused us accommodation.

We returned the way we came until we found the off-ramp that led to the motel we saw earlier. By now we felt anxious, not knowing if we would get a room for the night. The Privilege was painted a bright burned orange colour, with two rows of rooms on either side of a broadly paved inner driveway. One side had regular looking motel rooms for the likes of us; even a little patch of lawn at the end of the row, perfect for the dogs. The other side had drive in garages with automatic doors and rooms above, obviously for those who wished to remain incognito. This was verified by the tariffs posted on a board at the entrance, indicating that one could book a room for a minimum of five hours or by the night. This was definitely upscale, compared to the night before.

Privilege Motel tariffs

Hugo stopped at the by-now-familiar little window through which one communicated in these types of establishments. They had room for us. It was only after they had taken our money and showed us where to park that Hugo informed them that we had *perros*. We decided it was best to inform them after we had paid as we assumed it less likely that they would return our money and tell us to leave. After a slight hesitation, the management consented, provided we did not let the dogs sleep on the bed. The Westies would of course sleep there, but I had sheets with which to cover motel property. Even though we had a fold up cart, there was much baggage to lug in from the trailer and we appreciated the staff coming to give us a hand.

It was with great relief that we entered a comfortable room with regular and clean bedding. The bed base was a concrete box so common in Central America, but the mattress was a much better quality than the previous night. We also had very welcomed air conditioning and wifi connection. The two mirrors on opposite walls were tastefully hung so that only contortionists would be able to view themselves in any other position but upright. We fed the dogs and ourselves from the cool box while having thoughts of a juicy steak. We made ourselves comfortable for the night and opened our laptops to connect with the outside world.

Shortly after Hugo switched his on, it went black. Nothing he did could resurrect it from what I called, "Death by Virus". I gleefully reiterated the superiority of my Mac Book, which he now had to rely on.

On Thursday morning we were packed and on our way fairly early. From here we would be heading northwest, away from the coast and towards central Mexico. This time Hugo followed the GPS instructions to find our way out of the city, but we got hopelessly lost. We realised that even though it directed well on the toll roads, the GPS seemed totally confused in the cities. After an hour of unintentionally exploring the suburbs of Coatzacoalcos, locals pointed us in the right direction to the highway.

We could tell we were leaving the coastal plains as the road started climbing the gentle hills towards Cordoba. It seemed as if the Nissan also felt the climb, as Hugo announced that the vehicle all of a sudden seemed sluggish. I had visions of the vehicle breaking down in the middle of nowhere, with four dogs in a not-so-pet-friendly part of the world. To be on the safe side, we decided to drive into Cordoba and have the Nissan checked out. The first sign we saw on entering the outskirts of town read *Transmicion Mechanico*. In his broken Spanish Hugo was able to convey our problem, so with the trailer still attached and the dogs and I in the car, a mechanic took us for a test drive. I took note of a nice looking motel we passed in case we did have a breakdown, all the while being petrified that the trailer would come unhitched over the *topes* while Hugo was out of sight.

After the mechanic declared that he could not detect any problems and Hugo deciding that it was all due to him not taking the car out of cruise control up the hills, we resumed our journey. We passed a few more tollbooths before we reached the Rio Blanco Canyon National Park and the town of Nogales, where the road started to climb to the central plateau. We prayed that our vehicle would survive the extreme steepness, switchbacks and the many sharp and dangerous turns. We, along with many transport lorries, laboured our way up. The scenery was quite spectacular and different to the coastal plain we had just left behind: with relief we safely reached the top at an elevation of 7950 feet, and headed for Puebla.

In our original plan we would have stayed overnight in Puebla. That pleased me as I wanted to acquire a piece of Talavera pottery that Puebla is famous for. Now we arrived at midday and due to roadwork on the bypassing toll road, we found ourselves in a real mess of traffic. Hugo offered to venture into the city in search of pottery, but I could not face more bumper-to-bumper traffic while towing a trailer and thus adding to our stress, so we stayed in the jam and slowly inched our way forward. All was not lost; I had the name of a company in Tucson, Arizona, which was on our route, that advertised on the Internet that they imported Talavera from Mexico.

Beyond Puebla we had a few more checkpoint searches by both the army and the *federales* (federal police), but we encountered no hostility or hints of the latter requiring *mordida*. Inja seemed to dislike their black uniforms and made that known. We were however, still apprehensive about the environs of Mexico City.

As we journeyed on following the toll roads directed by the GPS, the tollbooths and tariffs increased in number and cost, but we did not complain as the roads were in excellent condition and we were awarded the safety provided by the many commercial transport lorries and other traffic on the road. The only mishap we came across was a burning lorry, which backed up traffic for miles, and fortunately only in the dual lane going in the opposite direction. Mexico City seemed to be nowhere in sight. It eventually dawned on us that we were on a new *autopista* that was bypassing the city completely.

We could hardly believe our good fortune. Our only remaining concern was to find a town before dark. Apart from our safety and police roadblock apprehensions, it had been a good day so far: Millie had not vomited (and would not do so for the rest of our journey); we seemed to have left the *topes* behind so the trailer was no longer coming unhitched; and we had not encountered any more rain, which was a good thing as we had not yet found a tarpaulin. We relaxed and enjoyed the scenery flashing by while the last remaining tatters of Dewey's linoleum flew off the trailer behind us and into the hot Mexican sky.

***

By late afternoon, it appeared that our only option for the night was the city of Pachuca, the capital of Hidalgo state. It was about thirty miles off the highway in rather arid-looking countryside. On the outskirts of the city, we noticed three auto-motels side-by-side; all protected from the outside by high fortress-like walls. We felt certain to find accommodation in at least one of them. Our cool box needed replenishing so we continued into the city to find a *bodega* (shop). With his smattering of Spanish, Hugo was able to get directions to a supermarket.

It was still a very hot day, and since we had two sets of car keys, we locked the dogs in the car with the engine and A/C on, while we went shopping together. We stocked up on more cheese, fruit and a barbequed chicken. We also bought a bottle of whiskey. As only the empty boxes were displayed, one had to take the box to the cashier, who then brought out the real article. I supposed, like everywhere else, Mexico also had shoplifters, not only drug lords.

Returning the way we had come, for no particular reason we picked the ABmar Hotel Motel. It looked like a large white concrete fortress with splashes of pink and yellow on the walls. A few yards down the protected-from-sight driveway, we came to the usual registration window. The smile on Hugo's face conveyed that we had received accommodation. Once payment was received, a security boom was raised and we drove forward into a large open-air and clean concreted quadrangle, flanked on three sides by rows of automatic garage doors with rooms above. As the car and trailer could not both fit into the garage, the proprietor himself came out to show us where to unhitch the trailer. It was at that moment Hugo informed him we also had clean *perros*. I could tell by the scowl on his face that he was not too pleased, but our strategy had again paid off.

We decided to leave the car outside and push the trailer into the garage, as that would give us convenient access to what we needed without unpacking everything. The dogs also had more room to stretch their legs and even pee against a wheel if necessary, which we would of course flush away with the water we carried. There was not a blade of grass in this quadrangle and we would have to walk the dogs all the way outside to the wild grass at the Pemex petrol station next door.

To access the room from inside the garage, we had to walk up a steep and narrow fake marble flight of stairs on one side of the garage. There was not a railing on the stairs and would not have passed security regulations in America, but Mexicans were obviously not that litigious. Besides, who would want to reveal their illicit trysts if someone lost their footing? The Westies bounded up the stairs to the room and we followed. This was definitely the most upscale love motel so far. We had the option of air conditioning or opening a window looking onto the courtyard. As usual, the bed had a sarcophagus like concrete base, but with a substantial mattress and crisp bed linen with a red bedspread. Everything was clean and tastefully appointed. If one did not know beforehand, the only give-away of the motel's dual purpose were the complimentary condoms on the bedside tables and the pornography on the television. Judging by our first night in a sleazy motel, there was a class system even in these nefarious practises.

We noticed that Inja was still in the garage and would not come up the stairs. It occurred to us that he was not used to more than two steps and not used to heights. The hazardous open spaces and not being able to get a good grip on the slippery tiles all added to his fear. No amount of coaxing budged him so with loud protests on his part and a pinch collar for encouragement, I eventually dragged him up the stairs. All the tension and effort of the past three days seemed to suddenly overwhelm me with a severe attack of palpitations and light-headedness. After lying down for a while and then being administered a stiff tot of scotch by Hugo, I felt sufficiently revived.

The following morning, after having packed up the trailer and ready to go, Inja was still in the room and too afraid this time to come down. Getting him down was a bigger challenge than getting him up; he could slip and in the process we could all fall through that open space. After some futile attempts and Hugo showing his impatience, we slung his blanket under his belly and half carried, half dragged this 110lb dog down the narrow flight, all while he was howling dog murder. We made a hasty exit before the proprietor could come to investigate. I could not wait to get across the border.

Our aim was to reach Saltillo on Friday afternoon, a 881 kilometre trip, which would enable us to reach the border town of Nuevo Laredo (across from the American city of Laredo, Texas) fairly early on Saturday. Relaxing with the thought that we only had one more night left on Mexican soil, we settled back to enjoy the passing scenery; with interruptions of only one military inspection and many tollbooths, the latter of which we had by now lost count.

We found the scenery and topography of Mexico very interesting. We were impressed by miles and miles of large agricultural fields; some harvested by machines and others by individuals doing backbreaking hand picking. Just by reading the labels in the fresh produce sections of Costco and ones local supermarket, one knows that all this productivity is to feed North Americans, for not as much produce comes from California as it did in the past. They appear to have developed different environmental priorities.

Not only did we see extensive agriculture along the way, we also passed what seemed like acres of manufacturing plants bearing familiar appliance brand names, which are only a fraction of all the varied industrial activities taking place in many parts of Mexico. As a consumer, it was interesting to read that there is a growing sense in industry that "reshoring" in Mexico has many advantages over "offshoring" in China. The quality of most goods from China leave much to be desired, especially if one has experienced all their junk that ends up in unsuspecting Third World countries where people can't really afford to keep replacing the same items that don't last longer than a few months.

Along our route we passed many places of natural beauty and historical interest. Travelling with a pack of dogs in the heat of summer, we just had to keep going and missed most of what Mexico had to offer us as tourists. It is unfortunate that Mexico's reputation has been sullied by drug cartels. On the other hand, if America did not provide such a lucrative market, the situation might not be so dire.

***

We reached Saltillo quite late in the afternoon. Without a city map to guide us, and a confused GPS, it was hard to find our way, so we just followed what seemed to be the main through road. We passed only one upscale looking hotel-motel on our way through the centre of town. It did not seem possible to deploy our dog strategy at this place, and they refused us accommodation when Hugo asked.

We had one address of a pet friendly accommodation but we could not find it. We continued on and nearly at the outskirts of the city we saw a quaint little adobe-looking building with a sign reading “Hotel Hacienda Real Suites”. We drove into a small paved courtyard with regular looking motel rooms leading out onto a covered walkway with colourful flowerpots. We saw only one other vehicle, so we were hopeful in finding a room.

At the registration desk Hugo encountered a young man who couldn’t speak any English. He managed to convey that we needed a room and that we had *perros*. The young man indicated he had to make a phone call. He returned with the news that we could take the small dogs into the room but the *perro grande* had to remain in the car. That was definitely not an option for us. After Hugo indicated that we would look for another place, the young man again made a second phone call and returned with the news that we could take the big dog into the room if we paid a fee: we were elated.

Our suite was not quite the same as one would encounter in an American hotel suite; although a little worn, the bed and settee were normal; there was a little kitchenette with a small refrigerator; Internet service and a television without pornography; and all in all the best night of our journey through Mexico. I discarded the last pair of sandals that had just barely been holding together with “shoegoo”. The following morning we could see the elderly proprietor and his wife scowling through the office window as we loaded our dogs and baggage into the car. We had not quite exited the courtyard when we saw the wife hurry across to inspect the room we had just vacated. She would find it in the same condition as we had.

We left Saltillo, elated at the prospect of spending the next night on American soil. We had read that Monterrey was a dangerous place, so we were glad that the toll road bypassed the city and we had not encountered any checkpoints. We reached Nuevo Laredo at noon. Hugo decided that the GPS directions to the border crossing were wrong, so he turned in the opposite direction: we were again hopelessly lost. After many twists and turns we arrived at the long multiple-lane line-ups for the crossing. Hugo picked what seemed like the shortest lane, only to find ourselves in the diplomatic lane. We were directed out of this lane, which took us completely outside the line-up perimeter.

While Hugo stopped the car to deal with a brain freeze, I led the dogs on a potty break. Before long a friendly *Mexicano* who could tell we were in the wrong place, came to offer his assistance. For a small remuneration he offered to get us back into the line-up. Hugo was more than happy to oblige with all the *pesos* he had in his pocket. With a gate in the high perimeter fence opened and the traffic held up for us, we re-entered a lane not too far from the bridge's approach.

It was the weekend and the busiest time to cross the border, so we inched slowly forward in a string of six lanes on the bridge crossing the Rio Grande River. The two outside lanes held buses and trucks, so from a brief glimpse I had of the river, it did not appear to live up to its name: at least not there. Vendors walked in and out between vehicles, in a last attempt to sell cheap tourist trinkets and tamales.

Eventually we reached the American immigration checkpoint, where the sight of us, two dishevelled-looking retirees with four dogs in a dirty orange SUV towing an out-dated little wooden trailer, put an amused smile on the officer's face. After checking our passports and a cursory glance inside the trailer, he waved us on our way. At last we had arrived in the United States, happy and very thankful for having travelled through Mexico for four and a half days without any mishap.

***

We arrived in San Antonio, Texas, by mid-afternoon and checked in for two nights into the nearest pet friendly but rather grungy looking Motel 6. We needed a day to rest. We requested a ground level room but the motel was fully booked. It seemed all the Mexicans who had crossed the border when we did came to this same motel for the weekend. A room on the third floor was all that was avaliable, which meant a major haul of our baggage.

Motel 6, San Antonio

The motel rooms led onto outside walkways with metal railings, through which one could see through to the ground below. It was fortunate that the motel had an elevator as Inja balked at the open stairway resembling an oversized fire escape. I had to quietly coax him along the walkway by walking on the outside. He always attracted attention, and with people milling around everywhere, I tried to hide the fact that my big dog appeared to be a sissy, especially from our dubious looking neighbour who was smoking next to the railing and had a pit-bull in his room. It was evident that not everyone carried plastic baggies to pick up their doggie doo-doo off the grass.

The rest of the afternoon was spent doing a large load of accumulated laundry, and for supper we ate wilted salad and greasy chicken wings from a take-out menu. Nevertheless, we had a relaxed night on a more comfortable bed, with no more worries about unfriendly pet accommodation.

The first order of the day on Sunday was to buy a tarpaulin to cover the trailer, as rain was threatening. We chained the trailer to an adjacent power pole and went unencumbered in search of shops. After ten years in an underdeveloped country, Hugo felt a little like Rip Van Winkle, as there were many new things to see. He spent a long time in Best Buy, feeding his brain on the latest in technical gadgetry and eventually emerged with a new laptop computer. Next we stopped at a Wal-Mart, where I found a lot of never-seen-before ready-made gluten free food that I could load up our cool box with. That night we dined on salami, guacamole and hummus with tostitos.

Since we were now on more familiar ground we could make on-line hotel bookings in advance. Our plan was to drive west on the I-10 to Tucson, Arizona and then northward via the Grand Canyon. As Tucson was a two-day drive from San Antonio, we planned to overnight in Van Horn, Texas.

Driving to Van Horn the next day turned out to be very hot but thankfully not humid. The countryside was mostly flat and semi-arid. Suddenly Hugo noticed that the Nissan was burning up way more petrol than was normal and the gauge was almost on empty. A red light on the dashboard started flashing "Service Engine". There was no sign of life for miles around and we again had unpleasant visions of having a breakdown in the middle of nowhere. At least we felt we could cope better here than we could have in Mexico.

All we could hope for now was to reach a petrol station soon. When it seemed we only had fumes left, we noticed a rundown Exxon station just off the highway near a sign that pointed to Butterfield. Once we got there, we had to idle in the hot sun while two cars that had already filled up their tanks, blocked the pumps while the people went to the washroom, rearranged their cars, kids and dogs. We could only shake our heads at their thoughtlessness. Hugo eventually asked them to move away. Once back on the highway, I voiced my concern that I felt Hugo was driving too fast, especially with an ailing vehicle. We were both stressed and argumentative.

***

With one long straight main street and low buildings spread out on either side, Van Horn conjured up visions of horses hooves going clip-clop down the dusty road in the days of the Wild West, before paved roads and law and order. It was a tidy little town with an equally tidy little Motel 6. From there we contacted the Nissan dealership in Tucson to arrange to have the Xterra serviced the day after our arrival. At the same time we booked two nights at the La Quinta Hotel. We had supper from the cool box and went to bed tired, but relieved to know our vehicle would soon be looked at.

It was another hot day when we travelled to Tucson. The red light on the dashboard would not stop flashing but the fuel consumption seemed slightly better. Apart from potty breaks for the dogs and refuelling, our only other stop was at an interesting looking tourist trap called Bowlin's Continental Divide near Lordsburg, New Mexico. We bought some pure unadulterated desert honey and a few other preserves, but passed on the trinkets. Hugo was going to buy a large brimmed Western-style hat, until we saw that it was made in China. When we reached Tucson in the mid-afternoon, we first went to Borderlands Trading Company that advertised Mexican Talavera pottery.

The store was filled with brightly coloured pottery and some nice pieces of mesquite furniture. After looking at all the displays to decide which pieces would make nice gifts, on closer inspection I was dismayed to see that all that Mexican pottery was inferior imitation and not authentic Talavera. I was able to console myself with the lovely piece Arezoo had given me as a parting gift when they left Belize.

***

Tucson appeared to be a rather artistic city, with interesting adobe-like houses and cacti gardens. The La Quinta Hotel was situated in downtown Tucson. When checking in, with the manager hovering close by, the reception clerk informed us that their pet policy stipulated only two dogs per room so we would have to book two. Both rooms were on the ground floor, but one was outer-facing where we could park right outside the door. It was into this room that we unloaded all our baggage and settled in for the duration. The room was the largest thus far on our journey, and a small refrigerator ensured that our stay would be comfortable. We would surely be having that longed for steak dinner in Tucson, and it was with much glee that we were directed to the "Texas Roadhouse" just a block away.

While the dogs were locked in the idling car close by in the parking lot, we went inside for our first sit down dinner in eight days. We drooled over the menu that listed every cut of beef, pork, chicken or combinations thereof imaginable, as well as a few fish dishes and a myriad of salads and side orders. The steaks ranged from six-ounce sirloin, to 23-ounce porterhouse T-bone. No wonder obesity was on the rise. We relished our perfectly cooked eight-ounce sirloin steak dinners and retired for the night feeling satiated and content.

Early the following morning Hugo delivered the Xterra to the Nissan dealer. He waited quite a while before the manager returned with the not-so-surprising news that an extensive service was necessary and many parts and plugs needed replacing. Hugo was given a courtesy car and told to return the next morning. We whiled away the rest of the day walking the dogs, emailing friends and planning the rest of our trip northward. As we did not want to put the dogs in the courtesy car, Hugo drove to the Texas Roadhouse to request take-out steaks, where he was directed to join others in an area called "The Corral", supplied with a barrel full of peanuts to nibble at while they waited for their orders. We were of course happy to eat steak again.

As soon as they opened for business on Thursday morning, Hugo was back at the dealer to fetch our vehicle. They did not have good news. Even though they had done a thorough service and had their best technicians working on our car all the previous day, the computer still indicated problems, so he had to leave the vehicle for another day. On Friday the news was even worse. They could not guarantee that we would reach Canada in one piece. The general consensus was that after ten years in tropical humidity, the electrical harness had deteriorated. It would seem the worrisome noises we had experienced along the way, were the last gasps of a terminally ill Nissan, brought on by a harsh life in Belize.

After considering all options, it seemed our best course of action, albeit a painful one, was to buy a new Nissan Xterra. There wasn't a second hand one available at that moment and smaller Nissan models would not be able to pull the trailer and accommodate all the dogs. The purchase was therefore set in motion and all the business, legalities and money transfers required for purchasing a new vehicle in the United States and taking it into Canada were promptly and efficiently handled in one day. We were grateful this misfortune had befallen us in Tucson and we shuddered to think what would have happened if it had been in Mexico. Back at the La Quinta, I had to extend our reservation for another night. The manager was not around this time and the reception clerk was good enough to cancel the second room. I spent the afternoon browsing at a nearby shopping mall and we ate our last steak dinner from the Texas Roadhouse that evening.

***

With a brand new white Nissan, slightly larger and more powerful than our older model, we were on our way heading for Motel 6 in Flagstaff. We gradually left the semi-desert behind us and we started seeing hills covered with conifers and the weather in Flagstaff was decidedly cooler. We planned to travel along the south rim of the Grand Canyon on Sunday, therefore avoiding the Skywalk to the west. I was not sorry about that as I didn't feel the need to stand on a piece of glass projecting into space about 800 feet above the canyon floor. I was more excited about spending the night at Lake Powell on Glen Canyon. On one of my many previous flights between Belize and Canada, the sight of Glen Canyon from the air had fascinated me. It took me a long time to identify what I had seen, so the highlight of our whole trip for me was being able to actually visit the place on terra firma.

As we travelled eastward up the south rim of the canyon, we stopped at most of the observation points and took many photographs of the spectacular views as they appeared in so many different hues and colours. After leaving the canyon we drove north through a Navajo reservation and stopped to browse through their arts and crafts stalls, where they sold mostly hand made pottery and silver jewellery. The GPS was set for Lake Powell, which we hoped to reach early enough to still see some sights. Unexpectedly, instead we reached a dusty little settlement called Marble Canyon. At the petrol station Hugo was told that Lake Powell was at least two and a half hours away. Due to an earlier road closure and a confused GPS, we had missed the detour to Lake Powell. It was not an option for us to turn back and get there after dark, so Hugo had to phone and cancel our reservation. It is an understatement to say that I was disappointed. I also felt annoyed that we had planned our route only on the computer and relied on a GPS without the added benefit of a road map. We had to continue with the road we were on and hope for the best before dark. There had been a lot of "hoping" on this journey.

By late afternoon we found ourselves in Kaibab National Park to the north of Grand Canyon. We suddenly came upon an inviting looking little lodge called Jacob Lake Inn. There seemed to be a few rooms, a restaurant, a gift shop and a petrol station. Hugo went inside to enquire about a room and returned with good news; there were cabins behind the lodge where dogs were welcome. The quaint little cabins were nestled amongst the ponderosa pines, each with a small wooden deck looking out onto the woods. Fresh pinecones littered the ground and squirrels were running around, doing what squirrels do. This was a delightful stop, the best so far on our journey and Hugo received redemption.

Our cool box was still chilled with hummus, guacamole, ham, cheese and a bottle of wine, so we enjoyed an al fresco supper on the deck, looking out on nature that reminded us of Canada. The dogs enjoyed being out in the fresh air too and we let the three Westies off their leashes for a change.

On Monday morning we headed for Utah and planned to spend the night in Ogden. We had reached the final push towards Canada and Hugo and I were both looking forward to ending this journey. It had been an interesting adventure in itself, albeit with some stressful aspects. We also marvelled at the dogs' stoicism: they had never barked, yapped, whined, chewed or made messes along the way.

We were still on the road leading out of the National Park when a deer near the side of the road distracted Hugo. An Utahan driver in an old lorry in front of us was distracted too, but whereas he put on his brakes without brake lights, "Mr. Leadfoot" kept up his speed. I looked up to see us speeding full steam ahead into the back of a now almost stationary lorry. I let out a loud scream which brought Hugo back to reality and he slammed on his brakes; coming to a stop about six inches from the lorry and leaving hundreds of miles of rubber on the road. The new and more powerful Nissan had saved the day. As we passed the lorry, the hillbilly look-a-like did not give us as much as a glance. In a serious tone Hugo asked "Did you see what he looks like — a real "*hoender nek?*" (Afrikaans for chicken neck) as if that was a justification for nearly causing a major crash. That remark broke the tension and caused us to laugh hysterically.

***

We enjoyed the scenery until we reached the environs of Provo and Salt Lake City. We travelled for miles in a traffic jam on the Interstate Highway with a myriad of white Mormon Church steeples that dotted the suburban landscape on either side. We reiterated that we were not big city people. Eventually we left the worst traffic behind us and arrived in Ogden. It was then that we realised that not all Motels 6 are equal. This one was cigarette stinky; one frequented by truckers and in an adjacent parking lot some refrigerator trucks idled all night. The motel's redeeming feature was a large expanse of well-kept lawn for the dogs to sniff around on. When we awoke the next morning, our unpacked clothes, dog sheet and even our pyjamas were permeated with the smell of stale cigarette smoke. We were more than happy to move on from this motel.

***

We arrived in Baker City just a short distance within the northeast corner of Oregon. It was a neat little town nestled between big rolling hills. After a stop at a supermarket to replenish our food supply, we checked into a pet friendly Rodeway Inn, managed by a friendly East Indian couple from somewhere in Canada. Here we had the first Internet connection since Flagstaff, so we were able to book ahead at a Motel 6 in Everett, Washington for the following night. We wanted to pass beyond Seattle so that we could make an early start for the Canadian border and avoid being stuck in the early morning city traffic.

When we crossed the Oregon-Washington border the next morning, we were filled with a sense of joy and excitement as familiar vistas of the Pacific Northwest slowly unfolded ahead of us. We passed through the Yakima Valley ripe with fruit orchards and vineyards. The evergreen trees multiplied as we approached the mist-covered Cascade Mountains, allowing us a glimpse of snow-covered Mt. Rainier. Beyond the Snoqualmie Pass we reached Seattle, early enough to miss the afternoon rush hour traffic as we passed through. The GPS guided us to Motel 6 in Everett: we had reached the last night of our journey.

I did the last load of travelling laundry while Hugo cleaned out the car and trailer of everything extraneous, in preparation to enter Canada.

It was a bright clear sunny day on July 18, 2013 when we travelled the last 140 kilometres of our journey from Everett to the Canadian border at Peace Arch. We were truly thankful for having travelled without any mishap for 7500 kilometres in seventeen days. The Canadian flag fluttering in the light morning breeze seemed to be waving us a welcome back. In ten years we had come the full circle.

Endnotes:

1 Publisher: Preview Pub; 4th edition (January 1999)
ISBN-10: 1880862476
ISBN-13: 978-1880862476

2 www.belizeretirement.org

3 www.urbandictionary.com/define.php?term=gringo

4 www.mongabay.com/history/belize/
belize- health_and_welfare.html

5 BEL-08: Belize needs a water and sewage treatment/Biodiversity Report Award. Biodiversityreporting.org
www.amandala.com.bz/
news/90-belizeans-access-sewage-services/

6 travel.cnn.com/explorations/life/most-hated-cities-861160

7 en.wikipedia.org/wiki/Roads_in_Belize

8 Publisher:HarperOne:9.2.2012
ISBN-10: 0062076213

ISBN-13:978-0062076212
thegospelcoalition.org/blogs/tcg/2012/06/18/why-you-should-consider- cancelling-your-short-term-mission-trips/

9 www.facebook/BelmopanHumaneSociety

10 inhabitat.com/yale-students-discover-rare-plastic-eating-fungus

11 news.bbc.co.uk/1/hi/sci/tech/1402533.stm

12 edition.channel5belize.com/archives/85903

13 www.ict-pulse.com/2013/10/snapshot-update-caribbean-reading-2015- broadband-targets/

14 www.7newsbelize.com/sstory.php?nid=28555

15 www.belizeinvest.org.bz/industry-opportunities/medical-tourism/

16 www.marlashouseofhope.org
www.twaw.org/ministry-areas-orphanages-children-s-homes-marlas-house-of-hope

17 onin/hhhexpl.html

18 Beske,Melissa. “If Ih Noh Beat Mi Ih Noh Lov Mi” (If He Doesn’t Beat Me, He Doesn’t love Me): An Ethnographic Investigation of Intimate Partner Violence in Western Belize”. Phd diss., Tulane University 2012 - ***used with permission***

19 ctv3belizenews.com/index.php?option=com_content&view=article&id=3807:amandala-article-written-by-colin-bh-comes-under

20 www.dbzchild.org/uploads/docs/complete_pgmale_social_participation_and_violence_in_urban_belize_grand.pdf

21 www3.weforum.org/docs/WEF_Gender_Gap_Report_2013.pdf

22 www.state.gov/j/tip/rls/tiprpt/countries/2013/215400.htm
7newsbelize.com/sstory.php?nid=17655

23 www.state.gov/documents/organization/204638.pdf
www.belizetimes.
bz/2013/07/12-human-rights-report-on-child-sexual- violations/

24 www.refword.org/docid/51c2f3d651.html

25 www.ilo.org/ipecinfo/.../download.do?...

26 www.unodc.org

27 www.7newsbelize.com/printstory.php?func=print&nid

28 www.washintontimes.com/news/2012/jan16/tale-of-two-small-countries — ***Excerpt used with permission*** *

29 ***Poem by Jerry Larder used with permission***

CPSIA information can be obtained at www.ICGtesting.com
Printed in the USA
LVOW07s0416120115

422352LV00001B/43/P